T0270960

THE NILE AND ITS MASTERS: PAST, PRESENT, FUTURE
SOURCE OF HOPE AND ANGER

The Nile and its Masters: Past, Present, Future

Source of hope and anger

JEAN KERISEL

Emeritus Professor, Ecole Nationale des Ponts et Chaussées, Paris

Translated by Philip Cockle

A.A. BALKEMA / ROTTERDAM / BROOKFIELD / 2001

By the same author:
Albert Caquot, créateur et précurseur, Paris, Eyrolles, 1978.
Down to Earth: The invisible art of the builder, Rotterdam-Boston, Balkema, 1987.
La pyramide à travers les âges. *Art et religions,* Paris, Presses des Ponts et Chaussées, 1991.
Génie et démesure d'un pharaon: Khéops, Paris, Stock, 1996.

Authorization to photocopy items for internal or personal use, or the internal or personal use of specific clients, is granted by A.A.Balkema, Rotterdam, provided that the base fee of US$1.50 per copy, plus US$0.10 per page is paid directly to Copyright Clearance Center, 222 Rosewood Drive, Danvers, MA 01923, USA. For those organizations that have been granted a photocopy license by CCC, a separate system of payment has been arranged. The fee code for users of the Transactional Reporting Service is: 90 5809 343 3/00 US$1.50 + US$0.10.

Original text: *Le Nil: L'espoir et la colère – De la sagesse à la démesure*
© 1999 Presses des Ponts et Chaussées, Paris

English edition published by
A.A.Balkema, P.O.Box 1675, 3000 BR Rotterdam, Netherlands
Fax: +31.10.4135947; E-mail: balkema@balkema.nl; Internet site: http://www.balkema.nl

A.A.Balkema Publishers, 2252 Ridge Road, Brookfield, VT 05036-9704, USA
Fax: 802.276.3837; E-mail: info@ashgate.com

ISBN 90 5809 343 3

© 2001 A.A.Balkema, Rotterdam
Printed in the Netherlands

Contents

PART 2

CHAPTER 3 – THE NINETEENTH CENTURY:
PERSISTENCE OF THE PHARAONIC DREAM   91

The goddess Neith

> I do not know much about gods; but I think that the river
> Is a strong brown god – sullen, untamed and intractable,
> Patient to some degree, at first recognised as a frontier;
> Useful, untrustworthy, as a conveyor of commerce...
> Keeping his seasons and rages, destroyer, reminder
> Of what men choose to forget.

The opening lines of *The Dry Salvages*, from *Four Quartets* by T.S. Eliot

At Saïs[1] in the Nile delta there is a tradition that the following words were engraved on the statue of the goddess Neith: '*I am all, the past, the present and the future*'. Neith was the great creative force who bore within herself both feminine and masculine qualities – she was 'the father of fathers and the mother of mothers'. At Esna she shared a temple with Khnum, he who used his potter's wheel to shape gods and humans. Khnum and Neith had together 'uplifted the firmament' and given form to the world with the 'seven life-giving words'.

Statues sometimes speak to those who have the patience to question them. Whereas old people tend to be talkative and to repeat themselves, I was fascinated by the stories told by Neith. Her fabulous memory stretched back almost to the Big Bang; she knew all about the long history of Africa, about the tremendous forces that had split the continent and given birth to that great rift valley bordered with volcanoes within which man – and the Nile – were born.

She was so proud of the Nile; it came from so far away and reached its fulfilment in its delta close to her statue, as if it wished to pay tribute to her.

She clearly remembered the age of the great Pharaohs, who saw themselves as the rulers of the whole world. Narmer, the first Pharaoh, had worshipped her. But her moment of glory had not come until the Pharaohs of Saïs had made her the protectress of their capital; then she had become the principal goddess of the royal city, armed with her bow whose arrows repelled malefactors. One prayed to her thus:[2]

---

1. Saïs, a city on the lower Nile where Athena was worshipped, according to Strabo, 20.
2. S. Sauneron, 'Litanies de Neith, dans Esna VII', *L'écriture figurative dans les textes d'Esna*, I.F.A.O., Cairo, 1982, pp. 36-38.

To Neith, the male who made woman,
To Neith, the female who made man,
To Neith, father of fathers and mother of mothers,
born before what was destined to be born was born,
To Neith, who makes the waters flood when the time comes,
To Neith, the Shining Neith, at the portal of the valley...

With the great kings of Saïs came the rebirth of an empire and renewed splendour and prosperity in the twilight of her glory. She loved to speak of the great Psamtek I who, after driving out of Upper Nubia the black-skinned barbarians who had proclaimed themselves Pharaohs, had reunited the country and then opened it to foreign influences. And of Neko II who, twenty-five centuries before a stranger arrived from the north, had undertaken to connect, by means of small canals, the two seas, the sea with the red waters and the other vaster sea to the north. For foreigners, these two Pharaohs provided interpreters for the visit to her sanctuary; thus it was that Cambyses, the Persian conqueror, came there to pray as soon as he arrived in Egypt.

The Greeks, she told me, held her in great reverence. One of them, Herodotus of Halicarnassus, came to enquire about the 'Egyptian barbarians'. On his return home, the Greeks decided to name her Athena, symbol of reason and the mind, for Herodotus had travelled all over the country, visited her monuments and had all their inscriptions translated for him, thoroughly absorbing the wisdom of the Egyptian people. That Greek had been an extraordinary spirit interested in everything: he had seen into the mysteries surrounding the birth of that land, handiwork of the Nile, and had sensed that her statue reposed on the glorious remains of past civilizations; and he had also sensed the location of the far far distant springs that fed the river, so generous in its gifts.

The priests had spoken at length to Herodotus about the cruel genius Cheops but had said very little about the great Ramesses II, who had covered the soil of Egypt with gigantic statues and temples so that people would remember him. How strange it is that the most powerful of men set their hearts on becoming immortal!

But what really angered Neith was the extraordinary proliferation of gods in Egypt, most of them minor deities of little interest, who ought to have been drowned, like the ones in Mesopotamia at the time of a major flood. What had that god Ammon done to draw so much attention to himself, he who had been totally unknown at the time of the great pyramids. And that Ptah who claimed to have raised up the sky, or that Noun who said it was she, yes she, who had created the firmament!

How well Neith understood the horror of Moses for this pantheon of gods whose members were represented on earth by animals. Moses had fled from Egypt with the Hebrews and had asked them to unite with a single god, a god of love, always present but invisible, with no image and no need for an image, a certain Yahweh who bade man devote all his efforts to an upright conduct.

Neith had nourished the hope that all the descendants of the followers of this great alliance desired by Moses would set an example of tolerance. How disappointed she was to see the arch-fundamentalist Theodosius declare war on those he deemed heretics because they professed a faith different from his own. In his eyes, the Egyptians had become oppressors of the true faith, and so he had forbidden worship of the Egyptian gods, razed all the temples and massacred the priests.

She was very clear-sighted about the future, so I questioned her about the people of today. Were they better or worse? Worse, she replied; their behaviour has become so strange that it has taken a certain Freud to seek out the reason; that doctor has acknowledged my gift of clearsightedness for, in his house on that northern island where he had taken refuge, he kept my statue on his desk. He was a man of great independence of mind who, himself a Hebrew, said that Moses, the father of his religion, was not a member of the chosen people but a true Egyptian.[3]

In this land, she told me, the pretensions of man are unbounded: an enormous wall of stones has been built across the sacred river and, now that its course has been stopped, the delta is receding. She was quite certain that one day soon her statue would be submerged by salt water advancing over the land. Then she would have her revenge: the huge lake behind the dam would become sterile and would be filled to the brim by the sediments which used to bring riches to the plain.

Then would men see how vain it is to oppose the cycle of creation conceived by her: the generous floods of the past would return and with them the fertility of the delta. Then would her statue at Saïs be venerated once again as in the distant past.

The goddess told her story at length. All the great ones of coming centuries, she said, would be incurably inhabited by the obsession to perpetuate their memory through exorbitant works.

Today I sift through the notes I took and make ready to tell you of the pharaohs of every age.

---

3. Cf. Freud, *Moses and Monotheism.*

Introduction

'When you do not know where you are going, halt awhile and look back whence you have come.'
(African proverb)

When they were children the Pharaohs, just like us, had dreams and the urge to explore. After the mysterious inundation of the river had subsided they would see strange figures come down from Nubia, taller and darker of skin than themselves. Where did they come from? The young Pharaohs must have asked themselves questions about own their past: the African proverb above has always been true.

We must think far back in time and space, from the splitting of Africa, the first gushing of the Nile and the gradual expansion of the human race along its banks, in order to grasp how closely the past and the future of Egypt are bound up with the mighty river. To understand Egypt we must first understand the continuity of the Nile – in both time and space – as the river that nurtured humanity.

This book is the fruit of three other works and a voyage.

The first book, *The Mismeasure of Man* by the American palaeontologist Stephen Jay Gould, tells us that only one per cent of living species have survived; animals make up a tiny and recent branch of a very long evolution that began with bacteria, and it was from this branch that man evolved.

The second book was *Le Rêve de Lucy* (The dream of Lucy), by Yves Coppens, which brings out the full complexity of human evolution up to *homo sapiens sapiens*. It led me to wonder how, 'at the end of a long period of inert, then living, then thinking matter', Lucy, modern man, 'human cranes', pygmies and Pharaohs fitted in.

The third book was *Egypt's Road to Jerusalem*, by Boutros Boutros-Ghali, the Minister for Foreign Affairs of Egypt at the time of Sadat, who saw clearly that Egypt's next war would be fought over the water of the Nile.

The journey was the one I made down the small valley of an affluent of the Nile where, at the time of the Fourth Dynasty Pharaohs some forty-five centuries ago, the Egyptians built a small dam to tame the violent storm waters. To me, the design of that dam was a model of intelligence; but only a few days later I visited Sadd-el-Ali, the Aswan High Dam, and it seemed to me that that immense construction negated both the spirit and the reasoning of the ancient Pharaohs: was it really necessary to drown some twenty temples of Lower Nubia?

The Egyptian civilization stretches back over more than five thousand years with hardly any change in the territory it has occupied, a circumstance that is almost unique among ancient civilizations: the city states of Mesopotamia were federated but their union collapsed under the blows of repeated invasions; the civilization of the Indus Valley did not last; in part of what is now Mexico, Olmecs, Aztecs and Toltecs succeeded each other after relatively brief spells in control. The history of Egypt, with its continuity in time and space, is unique.

One of the first questions one is tempted to ask is how the Pharaohs felt about their African inheritance, and we shall see how they stooped to certain forms of racial prejudice, which engendered a form of self-centred thinking that still underlies the water policy of their successors in the valley.

We shall then describe how, in the Egyptian plain, a human society was gradually shaped by the seasonal rhythm of the Nile, and highlight certain aspects of their public works that have hitherto been given scant attention. By examining the logistics of the major construction projects undertaken by the great Pharaohs of the Old Kingdom, we try to show the extraordinary creativity and calculated boldness of the first Egyptians.

Since the memory of the Pharaohs is still alive in the plain of Egypt, we shall try to identify certain common traits, including a characteristic desire by many of them to surpass their predecessors.

Do the great Pharaohs of ancient Egypt have something in common with the leading figures who emerged at the time the country was seeking to free itself from foreign occupation in the nineteenth century? Literally, the word 'Pharaoh' comes from *per da*, meaning 'great house', the dwelling of a superhuman being who mediates between the gods, his brothers, and ordinary mortals. How many '*per da*' kings, emperors or presidents have dreamed of constructions that would measure up to the Pharaohs!

The history of the Pharaohs is gradually becoming clearer but still tends to be subjective. Here we try to analyze how the memory of the Pharaohs has been handed down through the ages engendering 'Egyptomania' and pharaonic ambitions in the nineteenth and twentieth centuries. It was pharaonic ambition that took hold of the man who built the canal through the Suez isthmus and of the man who later nationalized that canal. The building of the High Dam and the project for a second Nile parallel to the present river are other instances of the same kind. These two constructions have become a challenge to the other Nilotic nations: the third millennium will see a merciless war for water along the Nile which will lead sooner or later to a reversal of traditional alliances and the emergence of common interests.

Part 1

CHAPTER 1

The Nile, an ancestral waterway

'Africa is most of humanity by any proper genealogical definition; ... the non-African branch... can never be topologically more than a subsection within an African structure.'
Stephen Jay Gould, 1981.

In Africa, the historian Fernand Braudel liked to say, one loses one's sense of time. Because everything moves so slowly, our brain finds it hard to grasp the expanse of those very ancient times and the various stages in the long history of the continent. And yet we need to know that history to understand the subject we are about to explore.

Both man and the Nile are the product of a geological accident in Africa. Some 225 million years ago all the world's continents were united in a single landmass called Pangea,[1] surrounded by a single ocean called Tethys (Fig. 1). A crack in Pangea in the form of a gulf opening out towards the west foreshadows our Mediterranean Sea; it would take another 160 million years for the Atlantic ocean to come into being, separating Africa from South America; it was their interlocking forms that inspired Wegener's theory of continental drift, now universally accepted after decades of controversy. Some thirty million years ago, Africa collided with Asia, exerting immense pressure: Sinai was uplifted and the eastern end of the Mediterranean closed off (Fig. 2).

At the same time, Arabia is moving away from Africa (Fig. 3). However, while the northern part of Arabia remains attached to Africa by a narrow band, the southern part is pivoting around the northern rim of the isthmus. This has opened up a new sea, the Red Sea, under which the two continents are being drawn further and further apart by an oceanic subduction movement.[2] South Arabia is moving away from Nubia at a rate of two centimetres a year, that is nearly a kilometre since the time of the Pharaohs; in a few million years an ocean will lie between Nubia and South Arabia, once called Arabia Felix, the land of the legendary Queen of Sheba, where Pharaohs went to obtain myrrh. In the most recent episode, east Africa is splitting along the parallel lines shown in Figure 4, which represent narrow (less than 50 km wide) spreading zones called rifts. Beneath

---

1. From the Greek *pangea* meaning '*all the land*'.
2. Zones in which the ocean floor is expanding as a result of rising magma.

Figure 1. Pangea surrounded by the single ocean Tethys.

Figure 2. About 65 million years ago, Africa separated from America and the Mediteranean took shape.

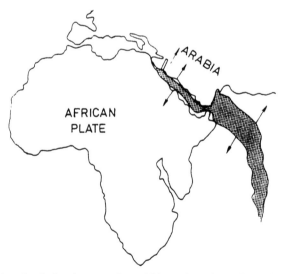

Figure 3. The Arabian plate is drawing away from Africa to form the Red Sea; the movement is much more rapid to the South, the Suez Isthmus acting as a pivot. The expanding ocean floor is shown in grey.

Figure 4. The rift valleys in East Africa are widening splits represented here by parallel lines.

these zones the earth's crust is thinner and at the edges of the rift hot lava-bearing flows are causing uplift and volcanic eruptions. It is these flows that have created the Ruwenzori range of mountains and Mount Kilimanjaro. Conversely, the zones of subsidence in the centre of the rift have led to the formation of lakes (Fig. 5). These rifts were destined to play a vital role in the birth of the Nile and the early history of man.

THE RIVER

BORN IN THE SOUTHERN HEMISPHERE, IN THE HEART OF THE RIFT VALLEY

The Rift Valley starts in the south, passes through Lake Tanganyika and, a little further north, divides into two branches, the more western one containing Lake Kivu, Lake George and Lake Albert while the other takes a north-easterly direction towards Kenya and Ethiopia as it leaves the Nile basin. The source of the Nile lies at 3°55'S, close to the western branch at an altitude of 2050 m. It was there that, on 12 November 1937, Burckhardt Waldecker discovered the ravine from which a thin trickle of water flowed, the real source of the great river; he set up a stela to mark the spot and on the nearest hilltop built a stone pyramid to remind people of the glorious destiny of the river which, during the inundation, would reflect the great pyramids in its calm waters (Fig. 6).

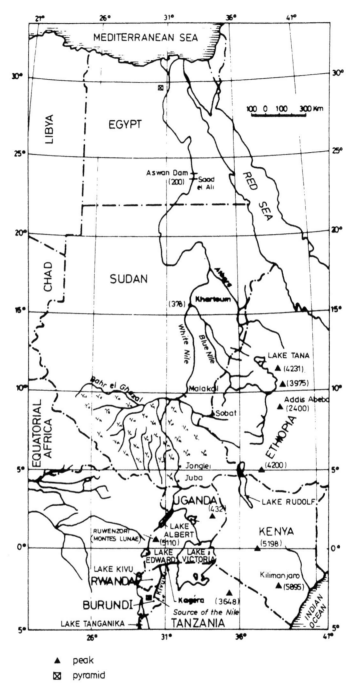

Figure 5. A series of lakes bordered by mountains lie in the rifts. The source of the Nile is in the southern hemisphere, marked on this map by a symbolic pyramid.

Figure 6. The pyramid erected at the source of the Nile is a replica of the Great Pyramid of Giza, which Waldecker had seen reflected in the Nile floodwaters (photograph by Félix Bonfils, 'View of the Pyramid of Cheops during the flood', taken around 1867-1870).

The exact location of the source of the Nile was for a long time so shrouded in mystery that in Rome there was a saying 'Quaerere fontes Nili' – to look for the sources of the Nile – which was equivalent to our 'to look for a needle in a haystack', i.e. to undertake something impossible.

The Egyptians placed the source *'in the land of ghosts and spirits'*, in the *'cavern of desires'*. When Herodotus investigated the subject, he did not rest content with the declarations of the scribe in charge of the sacred treasure of Saïs, who had told him (Book II, 28) that *'there were two mountains of conical shape called Crophi and Mophi and that the springs of the Nile, which were of fathomless depth, flowed out from between them'*. These mountains were said to be situated between Syena and Elephantine, that is to say at the northern point of entry into Nubia. He did not believe that information and asserted that *'the course of the Nile [extends]... as much further southward as one can travel by land or water in four months'* and even added that *'beyond, nobody knows its course with any certainty'*. A little later Aristotle simply mentioned that the river flowed down from *'a mountain of silver'*.

Cone projection

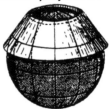

Figure 7. Ptolemy's conical projection. Ptolemy drew a map of the world on the basis of a cone touching the terrestrial sphere at 36°N, the latitude of the island of Rhodes.

It was the Greek astronomer and geographer Ptolemy (170-100 BC) from the School of Alexandria who made the biggest advance. Up until his time scholars depended on the very inaccurate geographical system established by Eratosthenes and the first maps of the world given in his *Geographia*. Using information gathered from the leaders of caravans returning from the heart of Africa, Ptolemy mapped the course of the Nile and the large Rift Valley lakes, placing the sources of the river at the foot of mountains he called the 'Mountains of the Moon', snowcapped peaks that he situated at latitude 12°30′S. In other words, Ptolemy gave the Nile, which was already very long, an extra thousand kilometres; moreover, the distortion in his conical projection (Fig. 7) aggravated any errors as one went south. The 'Mountains of the Moon' are in fact the Ruwenzori chain along the western edge of the rift and are located half a degree north of the equator; as we have said, the real source of the Nile lies some four degrees further south.

The high Ruwenzori (5110 m) is indeed a mountain of silver as Aristotle suspected, but the snows that cover it are lost in the clouds and seldom seen. But Ptolemy's *montes lunae* – mountains of the moon – continued to fascinate the Romans. Lucan has Julius Cesar say: '*No object has more excited my spirit: tell me what spring feeds the famous river, show me the place whence issues since the origins of time the long series of yearly inundations*'.

The next step was taken in the second half of the nineteenth century by explorers on behalf of the *Royal Geographical Society*. In its vision of Africa, imperial Britain saw its role as propagating the three C's – Civilization, Commerce and Christianity. Odd mixture of aims! Among the most brilliant British explorers were John Speke and Henry Morton Stanley. Speke discovered the northern exit of the Nile from Lake Victoria and, further upstream, the mouth of the Kagera, which he did not explore. This was done by Stanley later. Meanwhile the missionary Livingstone, perhaps led astray by Ptolemy's map, had mistakenly situated the source of the Nile much too much further south, near the shores of Lake Tanganyika. It was there that Stanley found him after a long journey fraught with danger and accosted him with the celebrated words 'Doctor Livingstone, I presume'. The famous missionary died during his expedition, after having tried to convert the local populations: he had some success but his teachings have since

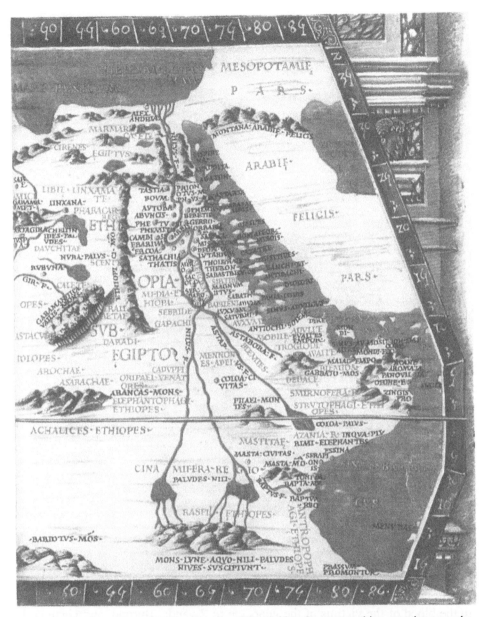

Figure 8. Fourth and last of Ptolemy's maps of East Africa. One can see his uncertainty over the sources of the Nile and his inscription 'Mons lune aquo Nili paludes nives suscipiunt'. It is also interesting to note the twin streams flowing through Nubia on either side of Meroe, suggesting that the uplifting of Nubia had ended quite recently.

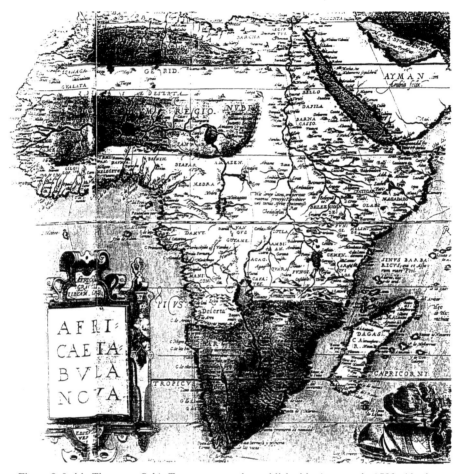

Figure 9. In his *Theatrum Orbis Terrarum*, an atlas published in Antwerp in 1593, Abraham Ortelius gives an even more erroneous representation of the Nile, making it flow through almost the whole of Africa from south to north.

been distorted and today almost all the local population are animist Christians. Fortunately, the reverend doctor died before the waters of the river were stained with the blood of the victims of the genocide in Rwanda, Burundi and Uganda. It was not until the twentieth century that the German Burckhardt Waldecker put the finishing touch to the long and fascinating search for the sources of the Nile.

THE ANCIENT NILES

The Nile as we know it is exceptional in every way: in length, in direction, in longitudinal profile and in regime it contradicts the geography of rivers. All these unusual features are bound up with the history of the river.

Scarcely 600,000 years ago the Nile, some 6000 km long today, did not exist. The area now covered by the Sudd swamps, in equatorial Africa, was an immense lake of over 200,000 km^2 which collected water draining down from the north (Fig. 10); north of Nubia, another quite distinct river, the Prenile,[3] flowed towards the Mediterranean.

The Prenile had a chequered history: some six million years ago, after a long period of glaciation, the level of the Mediterranean, which had become an isolated sea, dropped so much that it almost became a salt desert. The river, whose level at its mouth was that of the sea, cut into the limestone plateau (on which the pyramids would later be founded), gouging out a deep canyon as long as the Grand Canyon of the Colorado River. Later, the level of the Mediterranean rose and the Prenile canyon was filled up with sediments.

There were further rises and falls in the level of the Mediterranean, but much less extreme, so that the present-day Nile has cut a relatively small channel some 120 m deep in the sediment that filled the earlier canyon. The plain of Egypt is simply the product of another filling of the valley. In the words of Rushie Said, *'The Protonile was succeeded by two other rivers, the Prenile and the extant Neonile...the Neonile is a humble successor of the Prenile'*.

Between the Prenile and the Nile there was a whole series of Niles, none of which was in any way connected to the present source in southern Africa. They were all born in the mountains bordering the Red Sea, in the northern slopes of Nubia or possibly in the south of the Libyan Desert. The exact route taken by all the ancient Niles to the north of the Nubian Arc is still a matter of conjecture, except in the case of the first of these rivers, the Prenile, whose sediments have been found to the north of the Tushka Depression,[4] which suggests that the Prenile linked up the oases of the Libyan desert before finally returning to the northern part of its present-day course.

The Tushka Depression (Fig. 12) is a little known area of low-lying land that today plays a very important role in the functioning of the Aswan High Dam and in the project for a new valley. It is another abcess in the battle for water throughout the length of the Nile. The site has a rich history: at its western end lies Nabta Playa (Fig. 11), home of the North Sahara civilization, one of the forgotten peoples of the desert about whom we shall speak later, while to the east is a stela that recalls the activities of Chephren in this now desert area; here, to quote the words of Herodotus, *'the destiny of men was like a revolving wheel'*.

---

3. This is the name used today in place of the 'Urnil' employed in 1921 by the German geologist Blanckenhorn.
4. Cf J. Ball, *Problems of the Libyan desert*, 1927, and R. Said, *The Geology of the Nile*, Balkema, 1990, Chapter 25, pp. 506-7.

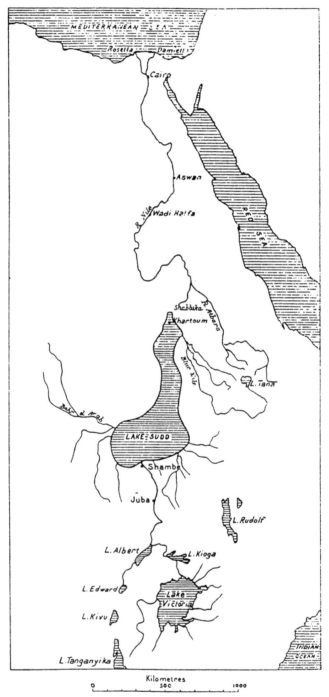

Figure 10. The two Niles: one million years ago there were two Niles flowing in opposite directions. The Blue Nile and the White Nile flowed towards the immense Lake Sudd while a Proto-Nile including the Atbara flowed northwards (from J. Ball, *Contributions to the geography of Egypt*, Cairo, Government Press, 1939, p. 77).

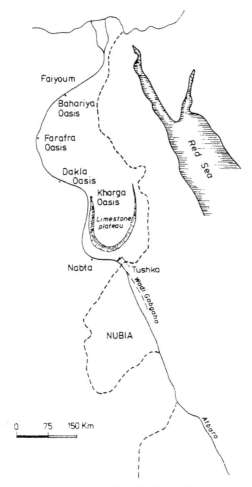

Figure 11. Possible route taken by the Prenile. The dotted lined represents the modern Nile that came into being after the uplifting of Nubia.

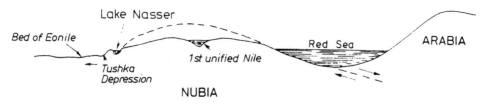

Figure 12. East-west section of the Nile before and after (shown by a dotted line) the uplifting of Nubia. The first full-length Nile flowed through Nubia by the same short route as the Prenile but the elevation of the Nubian formation some 125,000 years ago forced the Nile to find a way round it by the west.

TWO PECULIARITIES OF THE NILE

A singular profile

Normally, a river cuts its way from its source to the sea on a bed that, viewed from above, follows a single concave profile – a rapid descent from the mountains followed by a majestic progression in the plain. As we can see from Figure 13, the Nile is quite different: it proceeds in a number of steps, showing that it was once two separate rivers that have joined up.

The source of the Nile lies on the watershed where the streams on one side head south towards Lake Tanganyika while, from the other, the Nile boldly sets out northwards towards the tropical furnace. The tiny trickle of water that feeds the Rivuvu becomes the Kagera, which flows into Lake Victoria. From the other end of the lake the river plunges tumultuously towards Lake Kuga, then drops a further 40 m in an enormous cataract before it calms down and is reinforced by its affluent the Semiliki, which drains Lake Edward, Lake George and Lake Albert (formerly Lake Mobutu). At this point the river is called the White Nile. With a slope that is practically zero, the White Nile wends its way northwards towards Khartum for a distance equivalent to almost a third of its total length. At the beginning of this section, in the southwest of Sudan, lies the province of Bahr El Ghazal, an area that used to be well irrigated but is now suffering from famine.

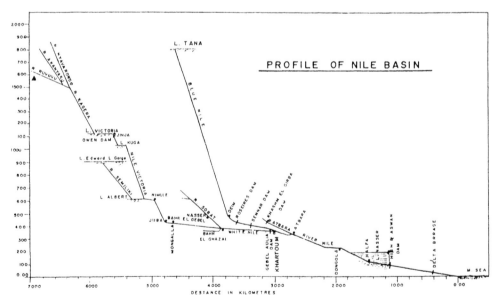

Figure 13. Longitudinal profile of the Nile from its source, the Rivuvu. Horizontal distances are in kilometres and elevation in metres (from A.B. Abulhoda in the International Commission on Large Dams (I.C.O.L.D.) in Cairo, November 1993, minutes p. 12).

From Khartum, swelled first by the Blue Nile and then by the Atbara, it makes its way round Nubia in a series of six cataracts. At the last but one of these cataracts, it has worn away the tender Nubian sandstone and makes its bed on a harder rock, forming a strange stony landscape that has since been drowned under the waters of Lake Nasser. The Nile at last arrives at Aswan and embarks peacefully on its last 1000 km or so towards the sea.

A singular pattern of flow

The discharge of a river is usually in proportion to the size of the water catchment area, but that is not the case with the Nile: a giant in terms of length it is a dwarf in terms of the amount of water it carries at the end of its course. With roughly the same size of water basin, the discharge of the Congo and the Mississippi are sixteen and seven times greater respectively. The White Nile loses up to two thirds of its water on its way through the Sudd swamps, the area of very slight slope mentioned above. In the Sudd numerous floating islands formed of a thick mass of aquatic vegetation function as dams, absorbing great quantities of water and slowing down the current. The enormous lakes traversed by the Nile earlier in its course are not enough to boost the outflow: they are like heated tanks in which almost all the rainfall evaporates.[5]

On the other hand the Blue Nile becomes, in midsummer, a collector of the enormous rainfall that occurs when the monsoon from the Indian Ocean meets the west wind from the Atlantic. This water rushes down narrow valleys, eroding the slopes and carrying off large quantities of sediment. The water arrives at Khartum with such force that, at the junction with the White Nile, it drives back the latter's current. Outside these four months of flooding each year, almost all the water of the Nile comes from the White Nile; this strange regime always astonished the ancient Egyptians.

Altogether, year after year, the amount of water carried by the Nile is 50-80% of Ethiopian origin and depends on that rendez-vous with the moisture-bearing winds over the Ethiopian highlands (see Figs 14 and 15). In other words, the great floods of the Nile are not born in the 'land of ghosts and lost spirits' or in the caverns of Crophi and Mophi; nor do they come from the melting snow of the Ruwenzori, the 'rainmaker'. The answer to the passionate questions asked by Lucan regarding the mysterious inundation is straightforward: the summer floods in the land of the Pharaohs are the handiwork of the monsoon that arrives in the Ethiopian highlands just before the summer solstice.

---

5. Present rainfall on Lake Victoria is around 100 billion m^3 per year, of which 94 are lost by evaporation; for Lake Kyoga the figures are 8 and 12 and for Lake Albert 3.6 and 6.3, i.e. evaporation is greater than rainfall.

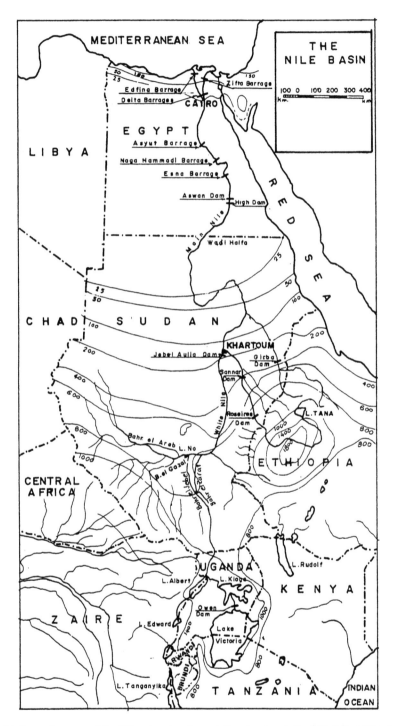

Figure 14. Present course of the Nile, showing the existing dams and pluviometric contours in millimetres of rainfall per year (from A.B. Abulhoda, I.C.O.L.D., Cairo, 1993).

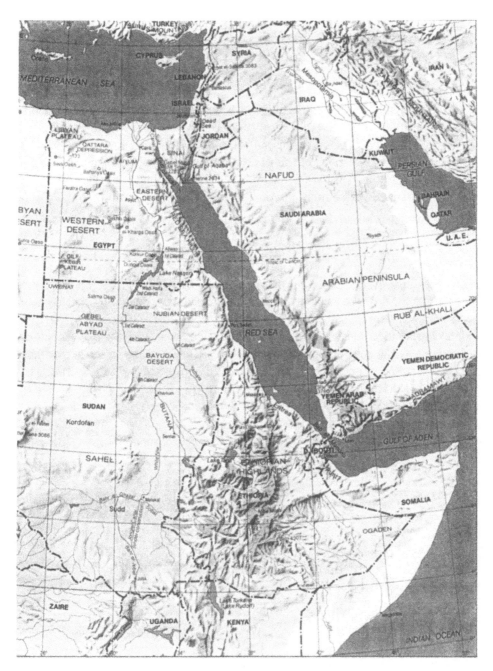

Figure 15. Relief of north-east Africa.

MANKIND

THE RIFT VALLEY: CRADLE OF THE NILE
AND OF THE HUMAN RACE

The history of hominids and subsequently of human beings, which occupies only the last ten million years or so, is closely connected with the formation of the rift valleys in Africa, which are like the scar of a immense caesarian operation on the African continent for the birth of the first hominids. Here many of the conditions were conducive to the evolution of primates – lakes and an equatorial forest bordered by a savannah with a more temperate climate during certain periods of glaciation.

Monkeys have existed for a very long time. Very recently, it was reported in the magazine *Nature* of 24 July 1997 that the skull of a monkey dating back fifteen million years had been found near the shores of Lake Victoria, a lake some fifteen times bigger than the Lake of Geneva and so named by the explorer John Speke in 1858. The most ancient primate so far discovered, this fossil has been named *Victoriapithecus*, suggesting an unexpected and rather comical comparison by an absent-minded and inadvertently impudent palaeontologist between a glorious Empress and a small monkey. Queen Victoria did not like to think that man descended from monkeys. The hidden dangers of compound words! Charles Darwin's very Victorian grandmother, who had once said to her grandson 'You may be descended from a monkey, but not me!', would have been highly amused.

This monkey, we are told by the palaeontologists, was the product of a branching evolutionary process reaching back to the development of bacteria (Fig. 16).

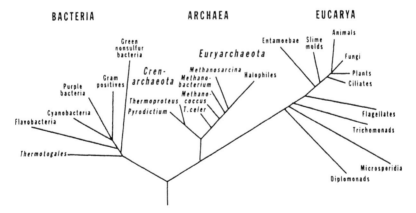

Figure 16. For some three billion years, bacteria were the only living organisms. This evolutionary tree, based on one by Stephen Jay Gould, shows that animals – including Victoriapithecus – are simply a branch of the Eucarya.

HUMANITY ON THE MOVE TOWARDS THE NILE VALLEY:
THE BRANCHES OF EVOLUTION

I have always been fascinated by the traces left by big mammals and early hominids: as I used to teach a discipline of geophysics concerned with soil mechanics
and the deformation of soils under the weight of heavy modern constructions, I
became increasingly interested in another more discreet science, ichnology (from
the Greek *ikhnos*, meaning trace), a branch of palaeontology that studies the fossil
footprints of large mammals. In 1974, while collecting photographs of the enormous footprints left in Bolivia by dinosaurs, I learned that the American palaeontologist Mary Leakey had recently discovered at Olduvai in Tanzania the footprints of two bipeds who had lived there about 3.5 million years ago. She very
kindly sent me a photograph of this historic discovery (Fig. 17). Long, long before our time, two small creatures (one about four feet six inches tall and the other
about four feet)[6] had walked side by side on a damp soil covered by a fine layer of

Figure 17. Olduvai – Mary Leakey beside the footprints of two hominids who lived around 3.5 million years ago.

---

6. Their height has been calculated by means of a rule, accepted by anthropologists, which states that a
   footprint measures 15% of the height. The skeleton of a hominid 1.22 m tall, who lived at about the
   same time as Lucy, has just been discovered at Sterkfontain, near Johannesburg.

ash from the nearby active volcano Sadiman. The footprints had then hardened. They are not the footprints of a primate: both the double arch of the sole of the foot and the curve of the big toe joined to the other toes show that these creatures walked like modern human beings. Hitherto our only knowledge of these distant bipeds had come from very skilful reconstruction, based partly on guesswork, from a few bones (tibias, backbones, etc.) and teeth; here we see them in action and our imagination wants to walk with them. They are humanity on its way to what is now Egypt and eventually to the four corners of the world.

These creatures were in fact two australopithecines, that is to say fossils with a number of simian features but already on the evolutionary road towards *homo habilis*.

Australopithecus diverged from the large primates around seven to ten million years ago[7] and became extinct some two million years ago. 'Lucy', who is regarded without good reason as the 'mythical grandmother' of humanity, was a little older than the hominids whose footprints were discovered by Mary Leakey: she was an australopithecine but still climbed trees. Lucy, named after a song by the Beatles, does not really deserve all the fuss made about her but one detail is worth noting: her brain cavity of around 500 cm^3 was ten times the size of that of *Victoriapithecus*. She had already travelled a long way along the evolutionary path since our own brain is only three times that of Lucy.

It was because the lithosphere was thinner in that part of the earth's crust that the Rift Valley came into being and created a moist and warm niche in which the monkey evolved towards the hominid and the hominid towards man; and it was also in this niche, as we have seen, that the great waterway began its long journey which would carry our common ancestors towards Egypt.

The pace of human evolution from *homo habilis* has varied over time; as before, new branches in the tree of life appeared, some of which just as suddenly expired, thus explaining the coexistence in the same region of advanced species of *homo* and surviving australopithecines. Lucy, despite the claims made by her discoverers, was most probably only one of our ancient great-aunts. There was crossbreeding and even specialists are finding it hard to unravel the strands linking, further back in time, *homo*, hominid, hominoid, australopithecine and pre-australopithecine. It is a forest of terms as thick as the tropical forest in which our distant ancestors lived.

It was this subtle jargon that led the French writer Vercors to publish in 1952 his novel *Les animaux dénaturés*, translated into English under the title *Borderline*. He imagined a young ethnologist working in the Rift Valley who comes across two female bipeds but cannot make up his mind whether they are 'homo' or 'hominid'. To decide he marries one of them. A child is born, which the ethnologist kills. He is brought to trial in London by a court which has to decide

---

7. This was the conclusion reached by a symposium held in the Vatican in 1980 by the *Pontifical Academy of Science*.

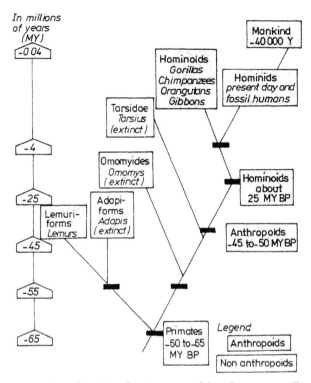

Figure 18. The most recent branches, since the emergence of the primates, according to Yves Coppens.

whether it is a case of infanticide or simply a hunting party. This satire about palaeontologists makes us feel uneasy but reminds us of the many instances of coexistence in the course of evolution.

Indeed, Lucy would have been truly astonished to learn that among her distant descendants could be numbered great Pharaohs as well as the strange beings found along the tropical part of the Nile, the 'crane-men' and the pygmies – pygmies in the virgin forest of the Democratic Republic of the Congo (formerly Zaire) between Lake Albert and Lake Edward and 'crane-men' (Fig. 19) further north in the islands of the desolate swamps known as the Sudd, where people still live as they did in Neolithic times. With their elongated head on a long body above long emaciated legs, they depend for survival on a small herd of cows.

How did this *homo habilis* of two million years ago evolve? His successor *homo erectus* appeared half a million years later but it is only 150,000 years ago that *homo sapiens* emerged. The final stage, *homo sapiens sapiens*, evolved at the beginning of the last glaciation, around 100,000 years ago and, however paradoxical it may appear, both geneticians and archaeologists date the movement out of Africa at this colder period. Setting out from the east of the continent, a tiny group of human beings gradually colonized the whole world.

Figure 19. A 'crane-man', still found even today on crossing the Sudd.

Small bands of hunter-gatherers took advantage of the low level of the seas to cross the Red Sea to Yemen and southeast Asia, while others made their way towards the northern Sahara and crossed the Mediterranean to settle in Europe. But the great majority followed the easier route represented by the Prenile. Step by step, from south to north, they took possession of the wide alluvial plain of Gezireh, in present-day Sudan, where the White Nile and the Blue Nile join forces to follow the short course of the Prenile through Nubia. The Nile plain had not yet been fully formed, so the most adventurous followed the eastern banks of the river, crossed the isthmus and settled the region that would come to be called the Fertile Crescent.

THE PLAIN OF EGYPT, HOMELAND FOR MODERN MAN
AFTER THE LAST GLACIATION

Migration remained on a small scale until the end of the last glaciation, around 15,000 years ago. As the ice melted *homo sapiens sapiens* settled what is now the Libyan Desert, in humid areas and oases. But during the periods of drought following the Ice Age, the deserts lost their inhabitants, who moved to the upper valley of the Nile at the edge of the future plain on terraces left by the river when the Mediterranean was, for once, at a higher level than it is today.

Climatic conditions were so favourable in the early Neolithic that certain ethnic groups settled and became peasants and stockbreeders, cultivating cereals and domesticating the donkey, the cow and other animals.

The last period of drought ended around 5000 BC and the historical record became less uncertain with the beautiful pottery placed beside the bodies in tombs. Their thermoluminescence[8] makes it possible to date fairly precisely during the fourth millennium the first triumphs of a technical revolution in agriculture at Nagada (site of El Amrah at the junction of the Nile and Wadi Hammamet) and at Gerzeh on a terrace near Faiyum. The richness of the tombs and funerary objects shows that these peoples attached great importance to their ancestors and that they had idols (Fig. 20).

Figure 20. Idol from Merimda, middle of fifth millennium BC (Cairo Museum). The holes in the head and face were used to attach the hair and beard.

---

8. Béatrice Midant-Reynes, 'L'Egypte prédynastique et les fouilles d'El-Adaima', in *Egypte 8*, p. 2.

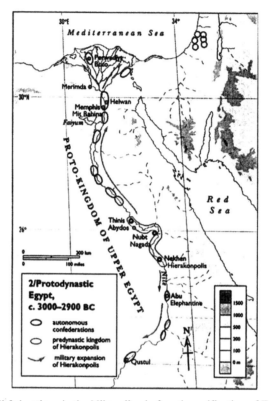

Figure 21. The small federations in the Nile valley before the unification of Egypt (from the *Historical Atlas of Ancient Egypt* by Bill Manley, 1996).

Narmer, a Pharaoh of the First Dynasty, made Hierakonpolis his capital. It was situated in an ecological niche where a wadi that no longer exists once joined the Nile. It seems to have been fairly prosperous, with some fifteen thousand inhabitants at its height. It was here that objects symbolising the new-found unity of the country, two clubs and a palette were found. The very length of the country made a strong central authority necessary: Narmer was the first king to exercise this power and it was during his reign that the small federations in the valley agreed to unite (Fig. 21), giving Egypt the beginnings of a backbone.

FORGOTTEN PEOPLES AND ANCIENT RAINS
IN THE SAHARA

We tend to concentrate our attention on the future plain of the Nile, neglecting the first settlements now buried under their shroud of dunes which shift hither and thither with each sandstorm in the western desert. But these regions, too, have their tenacious archaeologists. One of them is Fred Wendorf who, at Nabta Playa,

(Fig. 11), over a hundred kilometres west of Abu Simbel, has uncovered the remains of eighteen villages some nine thousand years old spread over an area of around 38 km[2]. At that time, this part of the Sahara was humid.

The most astonishing discovery was that of a series of upright stones[9] suggesting that these very early populations knew how to find the north and marked where the sun rose at the summer solstice. The stones form a kind of observatory that predates Stonehenge in England. In this region close to the Tropic of Cancer, three weeks before and again three weeks after the summer solstice, a perfectly upright megalith does not cast a shadow: the sun is then vertical and announces the beginning of the bountiful rainy season. Even in Neolithic times the first sedentary populations thus recorded the seasons. In that area and during that distant past, the sometimes generous and sometimes niggardly rains filled a number of seasonal lakes.[10] Large flat stones cover what appears to be a tomb and it is perhaps there that the very first Pharaoh was born and died.

THE PHARAOHS REJECT THEIR AFRICAN INHERITANCE

Since the time when Charles Darwin defined the principles of evolution and created such a scandal, there has been a big change in outlook. '*Evolution is no longer just a hypothesis*' recognized Pope John Paul II in his message of 24 October 1996. Nowadays almost everyone accepts a genealogy that reaches back to the apes, but fewer people accept that human skin was originally black and discussion continues about the conditions under which cultural evolution, a quite different problem, took place.

The ancestors of the Pharaohs were African and black-skinned. Even if we assume that hominids developed simultaneously in other parts of the world – such as China – it is difficult to imagine early humans coming from elsewhere to settle in Egypt during the Neolithic. And yet this question was a controversial issue for a long time; quite recently the Senegalese scholar Cheikh Anta Diop was still defending his theory of an African origin of the Egyptian civilization, though without much success.

It is true that an African Egypt was completely at odds with the affirmations of the highly respected English Egyptologist Flinders Petrie,[11] who argued that the population of Egypt at the time was the product of a white race that had migrated there from the Near East.

Further back in time, the French historian Constantin de Volney (1757-1820),

---

9. See *Nature* of 2 April 1998.
10. See the beautifully illustrated book by Pauline and Philippe de Flers, *Une autre Egypte, La Civilisation du Désert*, Paris, Mengès (in press).
11. See his two books, *The Races of Early Egypt*, 1901, and *Racial Photographs from Egyptian Monuments*, 1887.

who had sojourned in Egypt from 1783 to 1785, had recognized that the Egyptians were true negroes like all the indigenous peoples of Africa. He poured scorn on them but at the same time expressed astonishment at their knowledge. How was it possible, he wrote, ' *for this race of black people, today our slaves and the object of our scorn, to be that to which we owe our arts, our science and even our use of speech*'? This vigorous racism, long encouraged by the slave trade and colonialism, was alive and well on the eve of the French Revolution.

For two centuries the argument continued: were the Egyptians 'negro' or 'white'? In the hope of putting an end to this polemic, a conference was organized by Unesco in Cairo in 1974. Throughout the passionate debate the participants failed to agree on a definition for the word 'negro'. Eventually the Chairman, Maurice Glélé, a magnificent black from Cameroon, exclaimed ' *Ladies and gentlemen, what I want to know is – am I or am I not a negro?*'

Though no one wanted to admit it, two completely different issues – one the simple matter of skin colour and the other the much more complex issue of cultural evolution – were being confused. In Egypt, the physical type varies from Upper to Lower Egypt, as it tends to do in all countries that cover many degrees of latitude: skin colour varies from south to north since, in the long run, pigmentation depends on exposure to the sun;[12] there is no doubt that the Egyptians had a lighter skin than the Nubians, a point they liked to stress in their representations (Fig. 22). At the same time the profile becomes more graceful in less torrid latitudes.

Figure 22. Model of a body of Nubian archers found in the tomb of the prince of a nome at Assiut, dated at around 2400 BC (Cairo Museum). The Nubians were known for their skill in archery, which won them the nickname 'eye-spikers'.

---

12. For transpiration to play its role of regulating temperature, body hair must disappear; if a person is naked in a country with powerful sunlight, a screen of melanin forms in the skin which produces a bronzing that is deeper in colour the stronger the sun. It is a perfectly straightforward physiological process.

Figure 23. A Nubian woman as seen by artists of ancient Egypt: black curly hair, thick lips, black skin but not without grace.

As Jean Vercoutter stresses,[13] the first Pharaohs were not racially prejudiced: a royal prince of the Old Kingdom married a black woman whose features are typically negroid (Fig. 24). Vivant Denon (1747-1825), a French artist taken to Egypt by Napoleon, must have had this image in mind when he drew a Sphinx[14] with a negro profile. Moreover, since the representations of the Pharaoh Chephren display a certain resemblance to the great Sphinx, it has naturally been deduced that the Pharaoh was a negro.

In fact, the Pharaohs were always very curious about the people living far away in the south. The Pharaohs of the Old Kingdom were astonished to learn of the existence of pygmies, who were simply distant cousins.

We are in 2300 BC and Pepi II has just succeeded to the throne. Horkhuef, a high official of the Sixth Dynasty serving as a kind of Foreign Secretary, informs the Pharaoh that he has brought back a pygmy from an expedition to the distant south. This is what the king replies: '*You have told My Majesty that never before has a person of like mien been brought back by those who have journeyed through the land of Yam [...] Leave the others and take with you safe and sound the dwarf that you are bringing back from the land of the living horizon [...] to rejoice the heart of the King of Upper and Lower Egypt [...]. If he travels with you in the boat, place good men around him on both sides of the boat so that he will not fall into the water. When he sleeps during the night, place good men to sleep on all sides of him in his quarters. Check that all is well ten times a night. My Majesty wishes to see this dwarf more dearly than the fruit of the quarries of Punt'.*

---

13. J. Vercoutter, 'L'image du Noir en Egypte ancienne', *Bulletin de la Société française d'Egyptologie*, March 1996.
14. The Arabs called the Sphinx Abu al-Hol, 'Father Terror'.

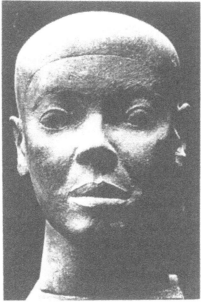

Figure 24. Head of a black princess of the Old Kingdom (W.S. Smith, HESPOK, Plate 6).

The number of precautions show how curious Pepy II, who is said to have lived for a hundred years, was about the African past and human evolution; he is like an ethnologist keenly interested in all the descendants of our dear Lucy. The Pharaohs of the first few dynasties were not racist and behaved with remarkable freedom: one of them, Neferkaré , became celebrated for a passionate love affair with his marshal. A papyrus well known to Egyptologists states: '*At night, the Pharaoh would withdraw and make his way to the home of his general Sasenet; he would throw a brick and have a ladder let down for him. He would spend several hours with the general . After His Majesty had done what he desired with the general, he would return to the palace*'.[15] Clearly, even forty-two centuries ago the local media were already taking an interest in what leading personalities were up to.

As dynasty followed dynasty, however, racial prejudice took root: on the tomb of the general Horemheb can be seen numerous images of negro prisoners of war crowded together and mistreated by Egyptian sergeants. As Jean Vercoutter writes, the African from Upper Nubia came to symbolize the vanquished – still a human being, but only just (Fig. 25).

---

15. See J. Van Dick, *The Nocturnal Wanderings of King Neferkaré*, 1994.

Figure 25. Head of a Nubian prisoner – End of Eighteenth Dynasty (Saint Louis Art Museum).

Egyptian women under the Ptolemies, like the Greek women, did not hesitate to daub themselves with ceruse (white lead) – despite its toxicity[16] – in order to bring out the whiteness of their skins.

Thus we see signs of spontaneous or deliberate prejudice towards these immigrants from the south or, to borrow the expression used by Claude Levi-Strauss, a kind of '*humanocentrisme*' which led to the ranking of certain populations as truly human and others as 'less fortunate'. And yet Darwin had told himself '*Never say superior or inferior*'. The Pharaohs found it hard to accept their forebears. The black Nubians from the Land of Kush eventually had their revenge during the first millennium BC, when a group of them conquered the whole of Egypt and became Pharaohs. Whereas the previous Pharaohs had spoken of their country as 'Kush the despicable', these Nubian kings showed great respect for the spiritual values of Egypt: a thousand years after the building of pyramids had stopped, they reintroduced the cult into their own country for their tombs.

16. Ceruse is now prohibited, even for artists.

The goddess Neith was therefore wrong to complain of the intolerance of Theodosius: her own Egypt had known periods of similar bigotry. Indeed, right up to the nineteenth century, Egyptians hunted blacks in the southern plains of the Nile with such determination that until quite recently slave traders hindered expeditions searching for the sources of the Nile.

On the other hand, the country adopted a very liberal attitude to peoples from the Near East. Assimilation was so successful that certain of the immigrants' gods were egyptianized: in the reign of Amenhotep II the Syrian and Palestinian gods Reshep and Astarte were worshipped; the Canaan god Hurun was associated with the Egyptian god Harmachis; and a number of Asians lived at the court of the Pharaohs of the New Kingdom, some of whom were appointed to important positions.

CHAPTER 2

The age of the Pharaohs:
Their subtle understanding of water

Towards 4500 BC began the long and glorious history of the Egypt of the Pharaohs, the mother of civilizations. We owe a lot to Rome, to Athens and to Jerusalem but, as the roots of those three cultures lie in Egypt, the study of that civilization is a key to modern humanism. It is a country in which dream and myth often
overlapped, a land of innumerable gods through which the Egyptians expressed
their wisdom and truths based on a rich metaphysic that already included such
major themes as eternal life,[1] the human god, the virgin mother and even the representation of a Trinity in the heavens.

The most fascinating pages of that history are the ones about the Old Kingdom,
when an almost human relationship grew up between a river and a civilization, a
river that would create a fertile plain and a civilization that would be regulated by
the rhythm of the world's most history-laden waterway.

LIGHT AND SHADE IN OUR KNOWLEDGE
OF ANCIENT EGYPT

Many writers have aspired to compile an exhaustive list of great human accom-
plishments since the beginning of recorded time. That was the ambition of Vitruvius when, during the reign of the Emperor Augustus, he wrote his *De Re Architectura,* a treatise copied and recopied many times over during the Middle Ages
and which the artists of the Renaissance regarded as the only important document
on the art of construction; such too was the ambition of Bernard Forest de Bélidor
with his *De l'Architecture Hydraulique.* Both Vitruvius and Bélidor believed, in
perfect good faith, that they had summarised the full range of human knowledge
in their respective fields; but neither of them had more than a smattering of what
had been done in Egypt. This was a most serious omission since it was the Pharaohs who opened the way in the art of construction and even more so in hydraulic

---

1. The symbol of the resurrection is found in the texts of the pyramids: 'O Pharaoh, you come out of your
   tomb as the Morning Star, you wend your way in the firmament and go to sit at the right hand of Ra'.

engineering. Auguste Choisy,[2] who was more aware of the achievements of the Egyptians, published his *L'art de bâtir chez les Egyptiens* in 1904, but his study needs to be brought up to date.

But how can we criticise Vitruvius and Bélidor when we think ourselves capable of describing faithfully the mastery of the Pharaohs despite the huge gaps in our knowledge. It is true that, over the last hundred and fifty years, the international cooperation of archaeologists has added greatly to our knowledge. But numerous aspects of life in the Old Kingdom are still cast in shadow: the papyruses of that period are few[3] and the ones that have survived form but a tiny proportion of those that existed. They have to be used with great caution as they are idealized biographies written for the gods, full of redundant expressions and boasting; the bas-reliefs too are idealized representations of their subjects.

Here as elsewhere, history has been a victim of the political context. Certain Pharaohs destroyed the works of their predecessors: Horemheb, for example, ordered his engravers to erase the names of Akhenaten and his two successors and replace them with his own; Thutmose III took his revenge on his aunt Hatshepsut, 'stepmother' and 'tyrannical usurper', by taking a hammer to her monuments and replacing her name with his own.

One is sometimes struck by an abuse of precision: the dates at which certain Pharaohs ascended the throne are given to the very year whereas the duration of the interregnums are unknown. Indeed, the latest datings by carbon-14 suggest that a number of Pharaohs about whom we know nothing at all need to be inserted at certain points. A stela contains the name of a king called Qahedjet who is not listed anywhere. Moreover, the lives of certain Pharaohs exceed human possibilities: one of them, Snofru, is said to have built no less than four pyramids, the last but one completed after the forty-seventh year of his reign[4] – and yet he left the reputation of having been a beneficent monarch. On the other hand, Pepy II, who is still spoken of as having lived a hundred years, appears, according to the latest research by Professor Jean Leclant, to have died at sixty.

The accounts we have are crowded with details about the lives of the Pharaohs, viziers and queens and about their rites and theology, but we know very little indeed about the economic and social organization of the country, about certain social and professional categories and especially about the daily lives of ordinary people, whose mummies show how grim their lives were. Lastly, we possess few genuine legal texts, so it is very hard to define the real extent of the royal power and understand the punishment of crime.

---

2. Choisy was an engineer and archaeologist often cited by the masters of architecture: Le Corbusier used to tell his pupils '*Read your Choisy*'.
3. The most ancient papyrus known goes back only to 2390 BC, whereas Narmer came on the scene around 3000 BC.
4. The duration of his reign is mentioned in the royal Turin Papyrus and the pyramids in question are Meidum, Seila, Dahshur South and Dahshur North.

Time and the wind have crumbled to dust the buildings of everyday life. As the harpist sang:

> 'But where are the mansions,
> Their walls have collapsed
> And even the trace of them has faded...'

Darwin once remarked that fossils are like a book of which only a few scattered pages, a few words and a few letters remain. All things considered, the history of ancient Egypt is a little like that. Some gigantic monuments have survived to the present day but do we really understand the country's infrastructure and the conditions under which the major works were constructed? Do we know all the subtleties of their management of water? – not so much the inundations with their attendant themes of fertility and renaissance so often mentioned in the records, but water diverted from the Nile and used as a highway for boats to carry materials for the great monuments at the foot of the western plateau. Where was Perunefer, the large Egyptian port at which vessels exceeding 50 m in length were built and repaired? Some of the scholars brought to Egypt by Bonaparte grasped the importance of studying all the hydraulic engineering systems of the past but did not have the time to go into detail.

Where were the walls of the fabulous Memphis located? That city, for centuries the beating heart of Egypt, had walls capable of withstanding exceptional floods, the force of which was increased by the obstacle in their path in the shape of the city itself. How high above the floodwaters were the ramparts of this immense city, long the biggest city in the world?

To answer such questions we need help from new sciences. The fertility of Egypt's plain depended not on myths but on physical factors pertaining to climatology (monsoon rains on the Ethiopian uplands) and geology (erosion of the Ethiopian slopes). The first advance was made in the nineteenth century by Boucher de Perthes, who introduced methods based on palaeontology and geology. At the time, the studies of this obscure customs officer in the small provincial town of Abbeville were treated with scepticism by the Academy of Science. But many sciences, long regarded as of minor importance by haughty archaeologists, have recently won recognition. K. Butzer, for example, in a study based largely on pollen analysis and geology applied to core samples extracted from the shores of African lakes, thought he had succeeded in reconstituting the climatic periods of ancient Egypt: 'It has become difficult,' he rightly affirmed, 'to ignore the possibility that major segments of ancient Egyptian history may be unintelligible without recourse to an ecological perspective.' This sound scientific ecology has given us a better understanding of the alternating periods of humidity and aridity at the dawn of the Egyptian civilization and of the period of drought that marked the end of its most brilliant phase.

Other sciences and techniques, such as biology, tribology (study of friction and lubrication) and the geosciences (especially hydrology and hydraulogy), are, as we shall see, casting an interesting light on the past. As these disciplines are based on fixed laws and logic, it is perfectly permissible to use them to add to our knowledge of ancient Egypt. It will not be easy; Pliny was quite right to say 'Ardua res vetustis novitatem dare' – it is a hard task to show the truth of what is old.

THE GIFT OF THE NILE

Let us go back fifteen thousand years – a split second of geological time. The most recent glaciation has just come to an end. There is no sea between England and France; the glaciers had stretched as far as the future city of Lutetia, now Paris. The Neanderthal is dead and his cousin the Cro-Magnon has taken refuge deep in caves in the flanks of profound valleys or not far from the future Massilia on the shores of the Mediterranean, a sea whose surface is about a 100 m below its present level; the Adriatic does not yet exist and Sicily is almost joined to Tunisia; in fact, there are two Mediterranean seas, the western one ending at the 'pillars of Hercules', the gateway towards the unknown, and the eastern one around whose shores the first great civilizations would develop. It is into this part that the Nile flows at the end of the deep gorge it has cut into the rock, for there is as yet no plain.

With the warming of the climate the Blue Nile and the Atbara hollow out their beds and channel immense quantities of water that tear fragments from the mountain slopes. *Homo sapiens* will soon start to settle on the successive levels of the valley bottom.

The subsoil of the plain of Lower Egypt is therefore of great archaeological interest. In that valley, as later all around the Mediterranean and on the shores of the Red Sea, where in Strabo's time there still lived troglodytic civilizations, there were many caves: the ones in the Nile gorge must have been like the Cousquer cave at Cassis on the other side of the Mediterranean; it is even possible that, as in the Cousquer cave, shamans capable while in a state of transe of entering into contact with the spirits left the negative imprints of their hands on the walls. The first Egyptians, too, painted silhouettes of animals to represent their gods. These impressions in stone revealed the first tremors of an artistic instinct; the Egyptian myths of the Duat, 'the underworld', were nourished by this memory of caves. There, each night, the Sun God traversed in succession the cave of life, the cave of Osiris, god of the Dead, and the city of deep waters and precipitous banks where he fought the serpent Apopis.

When Herodotus came to Egypt, he knew nothing at all about the history of the Nile but, as an excellent geographer, he was able to guess what had happened: '*Egypt itself was originally... [an] arm of the sea... running from the Mediterranean southwards towards Ethiopia*' (Book II, 11). His analysis is perfectly correct since, during the period of rapid warming after the last glaciation, the sea level rose more quickly than the deposition of sediments. Later the process of sedi-

Figure 26. The Sun God in his barque, sailing through the underworld (Book of Gates) (from A. Piankkoff and N. Rambova).

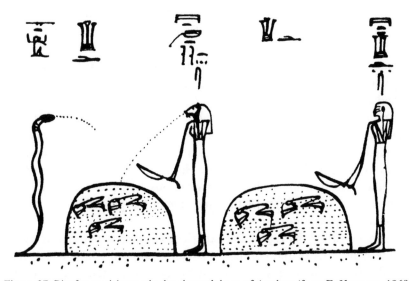

Figure 27. Pits for punishment in the eleventh hour of Amduat (from E. Hornung, 1968).

mentation gradually filled in the new gulf from south to north.[5] It is clearly the creation of this plain that inspired in the Egyptians their myth of emergence: the primordial water, the Noun, represents the 'non-being' prior to the moment of creation; from the mud of these waters is born the 'original mound' needed by the creator to carry out his work (Sarcophagus Texts dating from 2000 BC).

5. *'Surely in the vast stretch of time which has passed before I was born, a much bigger gulf than this could have been turned into dry land by the silt brought down by the Nile – for the Nile is a great river and does, in fact, work great changes,'* wrote Herodotus (Book II, 11) on his arrival, at a time when the plain had not yet been fully formed.

Herodotus, who was also an ethnologist, understood with remarkable perspicacity the irreversible south-north migration of the early humans since the depths of time: *'I do not believe that the Egyptians came into being at the same period as the Delta (as the Ionians call it); on the contrary, they have existed ever since men appeared on the earth, and as the Delta increased with the passage of time, many of them moved down into the new territory and many remained where they originally were'* (Book II, 15). He concluded that *Egypt is the gift of the river* (Book II, 5). Herodotus is talking about the subsoil of the plain; for him, the gift of the Nile is the creation of a national territory.

It has been argued that Herodotus stole his information from Hecataeus of Miletus,[6] another Greek who had visited Egypt in the previous century. I think this is mistaken; Hecataeus, the ancestor of the poets of antiquity, wrote *O generous Nile, you give life*, alluding to the fertilising sediment it carried, which is a quite different idea and, what is more, far from being always the case in ancient times. Indeed, I personally had the good fortune, when consulted after a major accident that occurred during the tunnelling of the first line of the future Cairo metro, to examine the data of numerous borings into the Cairo subsoil down to a depth of some 25 m. These soundings reveal an enormous disorder in the depositions of the Nile, with most of the material of no fertilising value at all since areas of stones, sand and silt are jumbled together.

It was in fact the finding of a shard of pottery deep in the subsoil that made me realize the full fury of the Nile's floodwaters which, when the river was younger, had destroyed the hopes of the first Egyptian settlements established in the valley bottom and laid waste their constructions. Hecataeus, ancestor of the poets, was dreaming: the Nile has not always been generous but, precisely because of its waywardness, it is the begetter of the land: far in the past, the Nile was both a *benediction and a misfortune*.

In this narrow valley the Egyptians experienced the destructive force of nature and become conscious of the conflict between order and chaos which the gods alone could resolve. On a canyon that had by then been totally filled in, ancient Egypt lived its hours of glory and was ruled by its truly great Pharaohs from Djoser to Ramesses III, including that demanding genius Cheops, the ostentatious Amenophis III, the reformer Akhenaten, the authoritarian and contested female Pharaoh Hatshepsut and the great Ramesses II.

If we believe Herodotus, the only one of these Pharaohs to conserve a memory of the past was Cheops, since his tomb, according to Herodotus, is a cave fed by the waters of the Nile: if this is true, it could be reasonably argued that Cheops wished to perpetuate the troglodyte traditions of his great ancestors.

---

6. Author of *Travels around the world* (Periegesis)

A HUMAN INSTITUTION FORGED BY GEOGRAPHY, HYDRAULICS AND ASTRONOMY

'Geography is the only invariable component of history', said Bismarck. This is particularly true of Egypt, whose somewhat absurd geography – a rectangle roughly 1000 km long by 10 km wide – has been the decisive factor. The vision of the orb of stars revolving around the north-south vector of water in a serene firmament gave the Egyptians the idea of gods inhabiting the heavens. '*Stars inhabited by individuals*', wrote Mahfouz, '*whom the Pharaohs turned into gods living light-years away*'. Moreover, the mystery of the distant sources of the Nile and the inability to explain the mechanism behind the flooding of the river, which followed a regular calendar but varied greatly in degree, must have nourished the image of divinity and the sense of eternity.

To give a firm basis for his power the Pharaoh upheld the idea that he himself was the intercessor between the here and now and the gods who govern the cosmos; the priests were subject to his commands and, at the bottom of the social pyramid, the humble farmers fed the nation. He saw society as a pyramid with a very broad base.

A number of ancient civilizations adopted the same image of a pyramid. The Mayas, for example, placed the *halac-huinic*, 'the one who knows', at the apex of the pyramid (Fig. 28): he would address only the five figures directly below him.

Figure 28. The pyramid of social organization among the Mayas (National Anthropological Museum, Mexico City).

Right at the bottom were the peasants, slaves and porters, the 'men of lower class'.

We shall now take a look at the social pyramid of ancient Egypt, commencing at the bottom with the fellahin, another forgotten people of history even though it was on them that the economy depended.

THE BASE OF THE SOCIAL PYRAMID: THE PEASANT, THE SOIL AND WATER

In the Republic of Athens, the citizens should, according to Socrates, be divided into three classes – rulers, auxiliaries and craftsmen. 'A stable society demands that these ranks be honoured and that citizens accept the status conferred upon them'. But how is their acquiescence to be secured? Socrates ends up by imagining the speech he would make to his fellow-citizens: 'Citizens, ... you are brothers, yet God has framed you differently. Some of you have the power of command, and in the composition of these he has mingled gold, wherefore also they have the greatest honour; others he has made of silver, to be auxiliaries; others again who are to be husbandmen and craftsmen he has composed of brass and iron; and the species will generally be preserved in the children.'

Even before Socrates, was it also divine intervention that fixed the rank of each Egyptian in the social pyramid, a rank that he would keep for ever? The peasants have never had an easy life in any civilization, but in Egypt they remained for thousands of years firmly bound to the soil with almost no hope of escape. From south to north, throughout the entire length of that narrow thousand-kilometre band, the toil of the peasant was always crushing and often desperate.

Natural embankments border the secondary bed of the river; these the peasants widened and raised to create the first farms. When the inundation arrives the water becomes reddish and assails these dykes *like a young man in love*, as the texts put it. At particular spots all along the river, cuttings are made to allow the silt-charged water to spread over the plain. Plutarch remarked that 'the waters of the flood mingle with the soil like blood with flesh'. Like a living body, the Nile is bled throughout its length, a kind of 'pelican that pierces its own flanks to feed its young'.

To preserve the blessing of this godsend, the peasants level the surface of the alluvial plain so that, when the inundation comes, the valley looks like a giant ricefield with basin after basin descending in tiny steps towards the sea. The operation of each basin is thus governed by that of its neighbours. Subordination, discipline, obedience, tenacity – and even resignation when the floodwaters are either insufficient or devastating – become the chief traits of the fellahin.

But that is not all that the peasant has to bear. The meanders of the river constantly change their course in the plain from inundation to inundation. At first, the settlements were located on raised embankments which became islands during the flood. Seeing this Herodotus wrote that '*When the Nile overflows, the whole country is converted into a sea, and the towns, which alone remain above water, look like the islands in the Aegean*'. A doubly dangerous situation: if the flood continued to rise, the fellahin would strive to build up the island and evacuate their herds to the edge of the desert before it was too late; if it rose still further it became each man for himself in boats and a major disaster, with ruin of the homesteads and destruction of foodstocks and seeds. But yet another danger lay in wait for the peasant: if, in changing its course, the river formed natural dams, the water would insidiously infiltrate the foundations of the embankments and then everything would have to be abandoned and a new settlement built elsewhere.

Thus, in this plain, in this aquatic universe where all is water, sediment, meander, rupture or reconstruction of dykes and ramparts, there exists a basic instability that constitutes a latent threat.

The fellah, who fed all the social classes of the society – priests, craftsmen, soldiers etc. – was a beast of burden at everyone's beck and call, under the orders of a 'scribe responsible for canals'. This scribe delimited the land to be sown with wheat, ordered the necessary public works (ditches, dykes, clearing of canals), requisitioned the necessary forced labour, estimated the proportion of the harvest to be levied, and sometimes even demanded the payment of taxes on crops that the peasant had been unable to sow or harvest. And when the King took a fancy to be munificent, the burden fell even more heavily on the peasants whether the floods had been good or bad, and their fate often lay in the hands of corrupt officials.

The bas-reliefs of certain tombs depict peasants being beaten on the orders of zealous officials. 'I had to obtain from my taxpayers 3632 jars of wine but I made them pay 25,368. I had to have 70 jars of honey and the honey I brought back filled 700 jars' we read in a report dating from the New Kingdom.

The peasant is treated with utter scorn by the grandees: 'He groans endlessly. His voice is as harsh as the cawing of a crow. His fingers and his arms suppurate and stink excessively. He is tired of standing in the mire dressed in rags and tatters' (From a satire on occupations composed between 2150 and 1750 BC).

It might be thought that the living conditions of the fellahin improved with time; not true. Here is what Geoffroy Saint-Hilaire, one of the scholars who travelled to Egypt with Bonaparte, has to say: 'Believe me, the great majority of villages are composed almost entirely of mud huts less than three feet high; the opening through which these unhappy creatures enter their hovels is a hole with a diameter of one foot and a half and this hole is always open; inside, there is only enough space for the husband, the wife and the children to lie squeezed against each other and, to enter their abode, they must crawl'. Later, we shall see that beatings remained common for a long time and were still carried out in the nineteenth century.

And yet one is struck by the gentle resignation of the Nile-dwellers, almost content in the belief that such extreme poverty does not entirely exclude a certain happiness.

THE MIDDLE CLASS: INFLUENCE OF THE
MILIEU AND CREATIVITY

The situation was different for the middle class: the life of a craftsman was less hard and he benefited from cultural progress. As Hans Goedicke[7] so rightly states, '*Ancient Egypt, throughout its long history presents itself as at least a two-tiered society, i.e. not all were socially equal, but an upper stratum and a lower stratum were distinguished*'.

How are we to account for the surprising creativity of the Egyptian artists? If we may hazard an explanation, their fecundity was the product of borrowings, additions or reworkings of very different traditions along the Nile from north to south, within a people with a diversified genetic heritage: skills, inventions and artistic expression are factors of technical and cultural progress. The Egyptian mastery was empirical and rudimentary: no brilliant individual genius but a long memory of successes and failures and the safeguarding on papyruses of the most precious knowledge. The unity of the Egyptian language was a considerable advantage but the few hundred words of the hieroglyphic writing system were not enough for the writing of major treatises: the memory was usually preserved by transmission through master craftsmen who travelled throughout the country.

Their brains functioned by a sort of 'neuronal Darwinism',[8] to use the expression of G. Edelman: certain circuits of no use to the artist would be abandoned in favour of connections repeated again and again during an apprenticeship. The potter at his wheel, the sculptor engraving hieroglyphs in stone – all these endlessly repeated operations become etched in the memory and leave indelible traces when coupled with a stimulating human environment, a harmonious landscape, a beautiful light, a rich metaphysic and, on top of all this, the patronage of the Pharaohs. The French historian Hippolyte Taine is criticised for being too simplistic when he shows that art depends on '*race, milieu and moment*'. We are shocked by the word 'race' because it cannot be defined scientifically. But what about 'milieu' and 'moment'? Do they not offer two plausible explanations in this case – that artists profiting from the psychological climate created by the Pharaohs were inspired by the plain of Egypt. Jean Grenier remarks that 'for each person there exist places that are predestined for happiness'. The valley of Egypt was predestined to be a centre of artistic creation.

What is particularly remarkable is that the Egyptian craftsman succeeded in expressing his artistic instinct by sculpting the hardest stones while the boat-builders succeeded in constructing the large fleets with timber of the poorest quality.

The craftsmen thus became more cultivated while the eyes of the fellahin never left the furrows made by their ploughs. When Egypt sank into the troubles of the

---

7. H. Goedicke, 'Water and tax' in *The institutional problems of water*, I.F.A.O. 1994.
8. G. Edelman, *Biologie de la conscience*, Paris, Odile Jacob, p. 199.

First Intermediate Period, the country lost its technological advance. New ideas came from countries to the east and the Egyptians took advantage of the techniques imported by the Hyksos dynasty (horses and chariots, bows and arrows, battle axes, etc.) but by the end of the Second Intermediate Period they had assimilated all this progress, which they would make use of to win back their independence.

THE DEEP UNDERSTANDING OF WATER IN ANCIENT EGYPT

MEMPHIS, THE GREAT CAPITAL WITH RAMPARTS ASSAILED BY THE INUNDATION

During the Fifth Dynasty pyramid-building declined and thoughts turned to the foundation of a great capital at the meeting-point of Upper and Lower Egypt. The Egyptians saw Memphis as a pivot of the 'Two Lands', a guarantee of equilibrium between the state to the north and the state to the south. The time had come for the monument, with its function as a religious symbol, to be succeeded by an expression of the power of a unified state.

In ancient Egypt, there were two types of city, those constructed on terraces on the banks of the river out of reach of the floods, and those mentioned by Herodotus as having been constructed on mounds in the valley. Memphis was an exception: it was located on the very floor of the valley and surrounded with ramparts that had to withstand the yearly inundation. It was a fortress called 'the White Wall' within which lay the sanctuary of the local divinity Ptah. For this reason it was also called the '*Castle of the ka of Ptah*'. Under the Sixth Dynasty it became the capital of the country and was given the name of Pepy I's pyramid, Mennefer, which the Greeks transformed phonetically into Memphis.

During the Middle Kingdom, the official capital became Thebes but Memphis, because of its location, surrendered none of its importance: under the Ptolemies, it even became the leading city again, arousing the jealousy of Thebes and, a little later, at the time when Herodotus was beginning his survey, it had the status of holy city and economic capital, arsenal of Egypt, centre of metal-working with arms manufacturers, ironsmiths and goldsmiths, and the principal starting-place for expeditions. As the great Egyptologist Gaston Maspero wrote, 'Memphis was for Greeks of the time what Cairo has been for those of today, the quintessential oriental city and eminent archetype of old Egypt. Despite the disasters that had befallen it in the preceding centuries, it was still very beautiful and, with Babylon, the largest city in the world. The religious festivals, especially those of Apis, attracted swarms of pilgrims at certain times of the year and bands of foreigners came from every possible race of the old continent to trade. Most of the nations that were regular visitors possessed their own quarter which took their name – the Phoenicians the Tyrian camp, the Hellens the Hellenic Wall and the Carians the Carian Wall – and there existed Caromemphites and Hellenomemphites in addition to the indigenous population. The White Wall surrounded a Persian army which the Satrap could call on in case of rebellion and the city could continue to resist long after the country had fallen into the hands of rebels.

'The city abounded in monuments: Temple of Phoenician Astarte where, since the Eighteenth Dynasty, priests from Syria celebrated the mysteries of the great goddess, Temple of Ra, Temple of Ammon, Temple of Atmu, Temple of Bastet, Temple of Isis. The still intact Temple of Ptah used to offer the visitor as admirable a spectacle as the Temple of Theban Ammon at Karnak. The kings had modified the original design according to their whim, one adding obelisks or colossi, another a pylon, yet another a pillared hall. As the work of some twenty dynasties, Memphis was like a compendium of the history of Egypt in which each image, each inscription, each statue attracted the attention of the curious visitor...'.

It was indeed to this city that Herodotus came to conduct his survey: 'There were other things, too, which I learnt at Memphis in conversation with the priests of Hephaistos (Ptah), and I actually went to Thebes and Heliopolis for the express purpose of finding out if the priests in those cities would agree in what they told me with the priests at Memphis.' (Book II, 3). For him, Memphis was the reference.

The foundation of Alexandria proved fatal to the economy of Memphis. Strabo, who visited Egypt in the second decade BC, speaks of a large and populous city, second only to Alexandria: '*Lakes stretch out before the city and the palaces which, now abandoned and in ruins, cover the top of a hill and descend as far as the lower city; nearby is a wood and a lake*' (32).

The prestige of the cults celebrated in Memphis, especially that of the bull Apis, remained high throughout the country. The living Apis, symbol of fertility and protector of Egypt's prosperity, continued even in Strabo's day to be honoured in his stable-sanctuary at Memphis in company with his mother, identified as Isis, and her numerous offspring.

Edrissi, writing in the twelfth century AD, would say that Memphis had become a field of ruins located around the small village of Mit Rahina. He was referring to an earth mound whose summit reached an elevation of 33 m, 14 m above the plain (see Fig. 29). In the fourteenth century the Mamelukes abandoned the site and the fabulous city soon disappeared entirely.

Where was the 'White Wall' of the fortress-city located? How were the ramparts constructed? The first of those two questions has always fascinated scholars but no one to my knowledge has tackled the second question, which concerns a matter that was crucial to the security of so great a city.

How difficult it is to bring to life a city that has been dismantled by limeburners and buried under the silt deposited by the floodwaters of the Nile and the sand carried by the winds of the desert. In the last ten years or so, however, the *Memphis Survey*, a systematic exploration by a British expedition, directed by D. Jeffreys and Lisa Giddy for archaeology and Professor J. Malek for epigraphy, has been carried out.

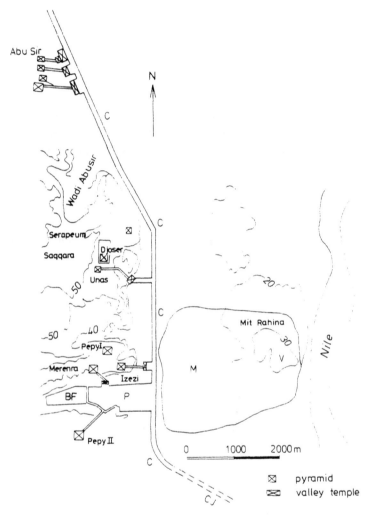

Figure 29. Presumed location of Greater Memphis (M), called Mennefer from the Fifth Dynasty, opposite the pyramid of Pepy I; location of the remains (V) near the present-day village of Mit Rahina.
Author's view regarding the location of Perunefer(P), the port of Memphis, with its wet dock (BF), the
lateral canal built by Cheops (C) and a possible route for the link canal towards the Nile (CJ).

An unreasonable risk

As the city grew in size, the whole undertaking proved increasingly unreasonable,
a challenge of water engineering, because it created an ever-growing obstacle to
the flow of the sacred river. Today people would regard the whole scheme as an
utterly senseless risk; indeed, our town councillors and planning authorities
rightly refuse to issue building permits in areas liable to flooding.

The Pharaoh's plan was audacious in two ways. Memphis lies upstream of the
delta in the narrowest part of the valley, which at that point is less than seven

kilometres wide, and the presence of the city would aggravate that bottleneck by taking up a third of its width. But the Egyptians had no choice: the desert was hostile and the site, at the junction of the 'Double Country', was in fact the only one capable of consolidating the political unity of Upper and Lower Egypt.

An enormous responsibility for the Pharaoh's architects

The architects and master-builders of the time had to find ingenious solutions. Failure was out of the question as the city was home to the Temple of Ptah, 'he who raises the firmament with his hand' and was addressed with the words 'Praise be to thee, Ptah, who contents the vital needs of men and gods'. Ptah was also the inventor of techniques and the protector of craftsmen, who had recently gone so far as to build houses of two storeys. Such a god was worthy of reverence and must be protected from the floods.

It was, of course, not always possible to keep up this unreasonable wager, as is shown by the layers of silt recently found by the *Memphis Survey* in the remains of the city.

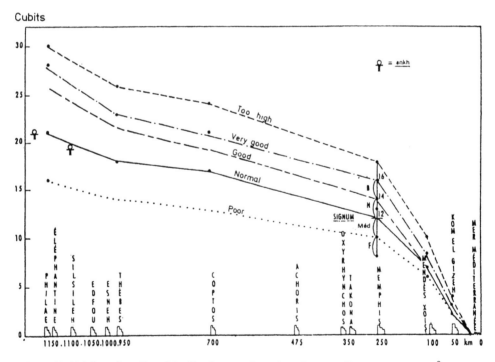

Figure 30. Height and quality of the floodwaters along the Nile, according to D. Bonneau.[9] A good inundation would attain 27 cubits at Elephantine (near Aswan) but only 14 at Memphis (1 cubit = 0.524 m.).

---

9 D. Bonneau, *Le fisc et le Nil*, 1971, Graph IV.

Advances in the art of building: The ramparts of the city

In Egypt, the art of building has a long history: there was a gradual advance from reeds to pisé, from sun-dried bricks to stone, from the heap of stones to the mastaba, and from the mastaba to the pyramid. The ramparts of Memphis reflected this progress.

Let us try to work out the various stages. What was the hydrological situation? Both Linant de Bellefonds[10] and D. Bonneau[11] agree that a very good flood at Memphis would be 8 m, while one of 10 m would be disastrous (cf. Fig. 30). The ramparts of the capital would therefore have to be eleven or 12 m high to protect the city from inundation.

How were the walls of Memphis able to withstand such enormous pressures of water? The problem was not to build a massive construction like a castle but to create a structure that would withstand the pressure of the water and at the same time prevent infiltration. There was a third danger: destruction of the foundations by the 'Noun', the name used by the Egyptians to designate the water table which, when the inundation begins, '*penetrates from below like underwater swimmers and steers the first floodwaters towards hollows in the ground*' (Aelius Aristides of Smyrna). To avoid the destruction of the foundations under the walls, the Pharaoh's architects decided to raise a little the ground-level within the city. This was perhaps the 'earthen beam' or 'earthen anchorage' mentioned in the texts[12]; the 'blocked cavities'[13] also mentioned could be balls of clay forming a sort of impermeable curtain to prevent the formation of cavities and the eventual destruction of the ramparts by the 'Noun'. The testimony of Strabo suggests that the most precious temples or palaces were built on a hill, of which the little mound of Mit Rahina is probably all that remains. The hill, in the centre of the city, was built up with imported soil, which explains the presence nearby of the lake mentioned by Strabo.

The first Memphis

It can be deduced from Herodotus's account that the city was initially situated close up against the flanks of the Libyan plateau: 'The priests told me that it was Min, the first king of Egypt, who raised the dam which created Memphis. The river used to flow along the base of the sandy hills on the Libyan border, and this monarch, by damming it up at the bend about a hundred furlongs south of Memphis, drained the original channel and diverted it to a new one half-way between the two lines of hills. To this day the elbow which the Nile forms here, where it is

---

10. Linant de Bellefonds, who became Linant Bey, Minister of Public Works under the Viceroys of the nineteenth century, observed the floods of the Nile for several decades. See pages 8-12 of his book.

11. D. Bonneau, *La Crue du Nil à travers mille ans d'histoire*, 1964.

12. J. Berlandini, 'Ptah-Démiurge', *Revue d'Egyptologie, 1995, Vol. 46, p. 14.*

13. Ibid., p. 31.

forced into its new channel, is most carefully watched by the Persians, who strengthen the dam every year; for should the river burst it, Memphis might be completely overwhelmed. On the land which had been drained by the diversion of the river, King Min built the city which is now called Memphis – it lies in the narrowest part of Egypt – and afterwards on the north and west of the town excavated a lake, communicating with the river, which itself protects it on the east. In addition to this the priests told me that he built there the large and very remarkable temple of Hephaestus' (Book II, 99).

This account is very interesting, as is often the case with Herodotus. The city was referred to as the 'White Wall'. As revealed by the findings of the *Memphis Survey*, the Memphis of Menes (Min for Herodotus) occupied a different site from the later city, which was located further north, near Abusir (see Figs 31 and 32).

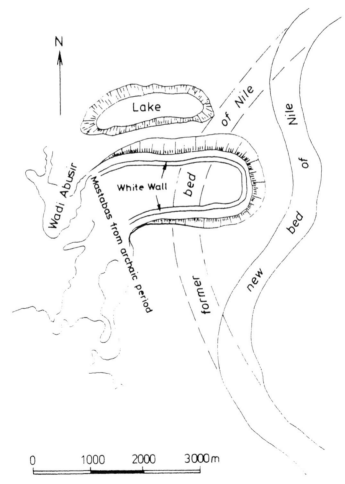

Figure 31. Site of the first Memphis and diverting of the Nile, according to the account left by Herodotus.

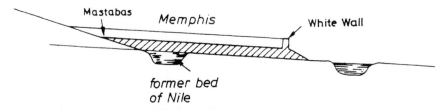

Figure 32. West-East section of the original Memphis with the first white wall, a kind of parapet.

Nearby, to the west, lie some mastabas of the pre-dynastic period. Traces of the lake that bordered the city to the north have also been found: it had been dug out by Menes to provide the soil with which to raise the level of his little city well above the level of the floodwaters. This artificial mound at the base of the hill had been surrounded, above the level of the floodwaters, by a small rampart of white stones from Tura[14] or white-painted bricks, which explains the name 'White Wall'.

The emblem of Lower Egypt was white: the city, founded by a king from that country, would necessarily recall that colour. The explanation offered by a commentator of Thucydides (Book I, 104), according to which the reference was to a great wall of white stones, must be rejected since the first large structures in stone were not built until the reign of Djoser, the builder of the stepped pyramid, some three hundred years after the founding of Memphis. Others have supposed that the reference is to the plaster covering of a wall made of sun-dried bricks: this theory too must be rejected because Egyptian bricks, like the gypsum of plaster, are vulnerable to erosion by flowing water.

The great Memphis, with its ramparts of dressed stone

In the beginning, therefore, the Egyptians concentrated on earthworks. But where did they find the enormous quantities of earth necessary for building the dykes of the large city that was to become Memphis? It seems that, as the city expanded, it was enclosed by walls of masonry. Instead of continuing to build gigantic pyramids, it was decided to create a large capital city upstream of Abusir. The Egyptians had by now mastered the art of cutting stone but, however perfect the joints, they knew that a dam-wall in stone would let the water ooze through the cracks and eventually bring down the wall. Water had its own laws to which man was obliged to adapt and any progress he made in bringing it under his control depended on an intelligent analysis of attempts that had failed.

The difficulty of the problem can be illustrated by the terrible accident that oc-

---

14. This does not contradict earlier statements: from the very earliest dynasties we find tombs of cut stone but cut stone was then a little-used novelty.

curred on the other side of the Red Sea, in Arabia Felix, in the kingdom of the Queen of Sheba, a queen who is severely criticised by the Koran: 'Her false gods have led her astray, for she comes from an unbelieving nation'. The capital of that kingdom was Marib, founded in about the eighth century BC. For nearly a thousand years Marib was an important centre for trade and agriculture. It owed its prosperity to a stone dam one kilometre long, behind which, during the rainy season, collected all the water from a catchment basin of some twenty thousand square kilometres. At one end of the dam, a system of sluices directed this water to irrigate a vast plain, transforming a sandy desert into gardens and orchards. The dam was the key to the power of the kingdom of Sheba. In 442 AD it collapsed under the pressure of the water, a disaster to which the Koran refers explicitly, telling of an angry Allah who 'unloosed upon them the waters of the dams... [and] punished them for their ingratitude'.

The photographs of the ruined dam (Figs 33 and 34) give some idea of the scale of the water engineering problem faced by the Egyptians when they constructed the ramparts of Memphis. They also show the beauty of the dam's original masonry.

Figure 33. Ruins of the eastern end of the Marib dam. The central part has vanished. A few inscriptions in Himyaritic attest that the dam was the pride and source of prosperity of the Sabeans.
(Photographs by Professor Soutif).

Figure 34. Inscriptions in Himyaritic capitals evoke the power and prosperity of the Sabeans.

This was not the only accident of its kind in history – such disasters occurred from time to time up until the end of the nineteenth century:[15] the skill of the stone-cutters has always led people to consider their work to be waterproof. As for the Egyptians, despite their supreme skill in the art of stone-cutting, they were well acquainted with the virtues and the vices of water. To avoid potentially destructive infiltration, they placed within the masonry an impermeable curtain of clay,[16] a kind of waterproof skin solidly fixed in position. The Egyptians observed nature with close attention to find solutions for their problems.

For certain external surfaces the white stone of Tura was used, giving the walls the appearance of an impregnable fortress mentioned by Piankhi the Nubian, who conquered Egypt and proclaimed himself Pharaoh in around 720 BC.

On the stela recounting the history of his conquest, we read that 'Memphis is fortified with a wall, a great rampart constructed with much cunning; it was impossible to attack it'. Each adviser then offered Piankhi his opinion. Let us besiege the place, said some; let us build a causeway, said others, let us make a

---

15. For example, there were two successive accidents with the masonry dam at Bouzey, in the Vosges, France, in 1884 and 1895. At the moment of its inauguration in 1884, the presumptuous chief engineer was giving a party on the banks of the reservoir.

16. This kind of waterproof inner core had already been used for the dam across the Garawi Wadi constructed during the Sixth Dynasty and walls of this type can be found, according to Dr Jeffreys, in the Fortress of Babylon in old Cairo.

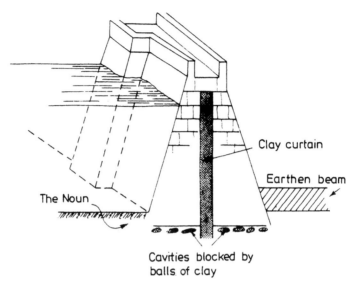

Figure 35. Possible section of the ramparts of greater Memphis, with the clay screen and the balls of clay under the foundations.

siege-tower; let us raise up a high mast; while still others, with less imagination suggested raising the level of the ground around. This testimony makes it clear that the structure in question was not an earthen dyke but masonry ramparts. We are reminded of a scene depicted on a bas-relief showing that, as early as the Sixth Dynasty, the Egyptians had to contend with the problems of besieging fortified towns (Fig. 36).

Piankhi got the better of the fortress which had such a mighty aspect by attacking it from the west, where he found a port. It was Perunefer, the port of Memphis. The gates of the city faced west and the canal of which we shall speak shortly passed between the western ramparts and the slopes of the plateau to give access to both the city on one side and the port dug out of the wadi on the other.

Size of Memphis

In its final state, Memphis stretched eastwards beyond the modern village of Mit Rahina, where the ruins of the Temple of Ptah have been discovered. From east to west the city covered about 2.8 km and took full advantage of the diverting of the Nile towards the east. As Lisa Giddy writes, 'the city followed the river'.

How far it extended in a north-south direction is a matter of conjecture. If we are to believe Diodorus of Sicily, the circumference of the city was 150 stadia, that is about 27 km, but he was a writer known for being insufficiently critical of his sources and prone to whimsical dating. It is hard to imagine the Egyptians building and maintaining such a length of ramparts, as a single metre of poorly

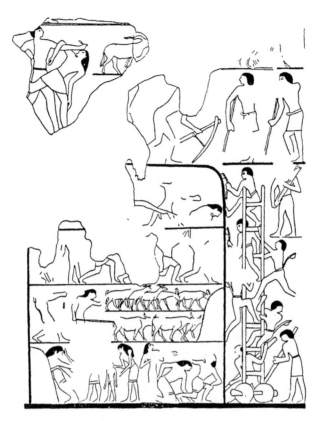

Figure 36. Siege-tower, a war-engine on wheels: relief of the Sixth Dynasty (from W.S. Smith, *History of Egyptian Sculpture and Painting in the Old Kingdom*, Figs 85 and 86).

constructed walling would cause the city to be flooded. Nevertheless, a number of archaeologists, including Flinders Petrie and François Jomard, have followed in his footsteps and taken the view that the city expanded northwards as far as the southern limits of present-day Cairo.

The cities that had preceded Memphis, such as Catal Hüyük in Anatolia and Jericho in Palestine, were tiny in comparison. Almost contemporaneous was Mohenjo-Daro, the great city of the Indus Valley, but its circumference was only 5 km. That of the Babylon of Nabucodonosor II, which so astonished Herodotus, was only 6 km, the same length as the walls of Paris under Philip Augustus in 1190; even the Paris of Charles V, who made the city of unrivalled importance, had a circumference of only 8 km in 1370.

Let us therefore leave Diodorus to his exaggerations and, basing ourselves on the *Memphis Survey*, the testimony of Piankhi and our own research in connection with the canal and the port, propose a reconstruction (Fig. 29) which gives the city a circumference of 11 km (2.5 km from west to east and 3 km from north to south), largely sufficient for the needs of this great religious, political, economic and social capital which has always fired the imagination. On the basis of these figures it is possible to calculate that the volume of masonry for the entire length of the ramparts was less than that of the Great Pyramid and that it undoubtedly lay within the possibilities of the Pharaohs.

But a number of unanswered questions remain concerning this fabulous city. Were the ramparts properly maintained during the Intermediate Periods? How did people manage during the inundations? And how, at that time, did the ambassadors who flocked there from all over the known world accede to the city? Did they arrive in boats that moored at one of the gateways whose threshold would be raised while the inundation lasted? Were there chains across the canal, as there were across the Seine in the time of Philip Augustus, to prevent the incursions of Libyans from the desert?

Access from the Nile: A waterway
to the foot of the western slopes

The ramparts of Memphis were a masterpiece in the art of construction, and the branch canal was another masterpiece in hydraulic engineering.

The wind has been the great destroyer of the hydraulic systems of the Pharaohs. Whenever the upkeep of the canals was neglected they were soon clogged by sandstorms from the west, which modified the general aspect of the plain. Here archaeology is powerless and the texts say little; when canals are mentioned, the hieroglyph *mr* does not distinguish between an irrigation channel and a waterway for boats. The nature of such a waterway has to be deduced from the logistics of the Pharaohs' projects.

There were many reasons for digging a lateral canal. The wheel was unknown and the only help that humans got for transport by land was from donkeys. The most practical means of transport was therefore by water, but the river also had its dangers since, once the flood had subsided, sandbanks would be found to have changed their positions and could wreck the heavily laden boats.

In the valley, between the mountains on either side, the river – as already noted – was tending to flow further and further way from the western slopes, where all the pyramids and all the temples of the religious communities were built. The purpose of the canal was therefore to supply the temples and pyramids by water. Some scholars think that, throughout the pyramid-building period, the Nile stuck like a limpet close to the flanks of the western hills. This is a lazy avoidance of the issue, for the meanders of a river, like an advancing worm, never cease to change their course as it progresses over a plain. Others have imagined a second river flowing at the foot of the plateau – a unique case in the annals of hydrology! As the laws of hydraulic engineering have not changed, it is not too presumptious to attempt to describe the principal features of this canal.

Earlier on, the building of the southernmost pyramids of Meidum and Dahshur had required the construction of canals leading westwards from the Nile to the

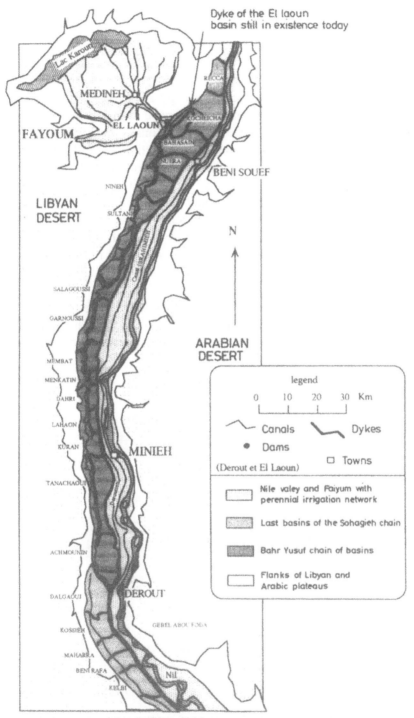

Figure 37. The series of basins of Bahr el-Youssef and the area of perennial irrigation by the el-Ibrahimieh Canal in around 1880 (from H. Barois, *L'irrigation en Egypte*, 1887).

foot of the Libyan hills so that the materials could be delivered without unloading and reloading – Tura limestone from further downstream across the river for Meidum and cedarwood from Lebanon in the case of Dahshur.[17]

The idea gradually arose of supplying the landing stages of the pyramids by means of a canal from Faiyum, the garden of Egypt, to the capital Memphis. Then Cheops decided to construct his enormous pyramid, standing like an outpost guarding the north, and so the canal was prolonged as far as the port of the Great Pyramid; this waterway then became busy with considerable traffic for the building of the two great pyramids and later for constructing the walls of Memphis and the other pyramids and for supplying the religious communities. Running parallel to the Nile, it became vital to the country's economic life and an important axis for the development of Lower Egypt.

During the inundation, the location and bends of the canal were indicated by beacons, at the top of which a piece of cloth would show the direction of the wind; the whole arrangement (Fig. 38) is expressed in the hieroglyph[18] for the divinity. The system of beacons indicated the invisible snags under the silty waters of the inundation.

It was the first large branch canal built by man, long before that of the Euphrates that Sargon, King of Assyria, had excavated to supply Mari and even longer before the thousands of branch canals dug to drive watermills in the Middle Ages and, in our own century, huge turbines.

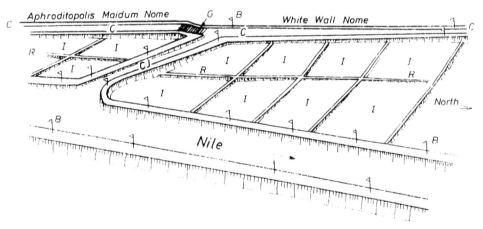

Figure 38. Diagram of the plain at low water time. At the top, the canal (C) and the slipway linking two nomes (G); the link canal from the Nile (CJ); irrigation channels (R) in the irrigation basins (I) and the beacons along the canals (B).

---

17. Some thick cedarwood props positioned as emergency measures in the damaged chambers are still to be seen.
18. See Appendix 1.

From nome to nome: the chain of water basins

Whereas the Nile glided smoothly towards the sea, the water of the irrigation basins on each side dropped from dyke to dyke in small steps (see Fig. 37) while further west, against the hillside, the canal made its way in a few reaches towards the sea (see Fig. 38).

The plain from Memphis to the site of the Great Pyramid formed a single entity, the nome (province) of the 'White Wall', while further south was the 'Aphroditopolis-Meidum' nome and beyond that the nome of Faiyum.

In each nome the water was perfectly horizontal, forming an unbroken expanse of water called a 'horizon'. That of the 'White Wall' nome was the horizon of Cheops (in today's nautical language these 'horizons' are called reaches). The successive stretches of level water, gradually dropping in steps from south to north, thus form a kind of staircase in which each step probably marked the boundary of a nome.

Passing from one 'horizon' to the next

How did the boats get from one horizon, or reach, to the next?[19] They were unloaded, hauled out of the water directly on to a lubricated slipway[20] and then refloated on the next reach. The idea of constructing slipways came to the Pharaohs when they tackled the difficult problem of getting past the cataracts on the Nile: to avoid the dangerous currents, the boats were unloaded and placed on a long slipway on the banks of the river, with forts to defend the pharaonic fleet from attack while it was out of the water. Jean Vercoutter has found the remains of one of these slipways bordering the second cataract at Mirgissa. They were the ancestor of the little slope used by millers in the Middle Ages to relaunch the boats that had delivered grain to the mill. They were also the ancestors of the slipways still used for the launching of liners and warships.[21]

At each change of nome, probably as far as the Mediterranean, a toll had to be paid. Diodorus reports the existence, at the end of the Bahr El-Youssef on leaving Faiyum, of a hydraulic construction to gain access to the Nile valley, at which a tax of thirty talents was levied. Proof of the existence of such tolls but also of two mistakes: the tolls could not possibly be as high as that and the tax must have been in kind since the Egyptian economy at the time was based on barter.

The biggest and longest reach was between Memphis and the Great Pyramid (Fig. 39): two boats were able to pass at the same time. Through an opening in the Nile embankment (Fig. 38) boats of all sizes, after lowering their sails (Fig. 40), would

---

19. It should be remembered that locks were not invented until the fifteenth century.
20. In England, in the Severn valley, many visitors interested in archaeology come to see the famous inclined plane by means of which barges attained the River Severn.
21. *Principles of Naval Architecture*, The Society of Naval Architects and Marine Engineers, 1967.

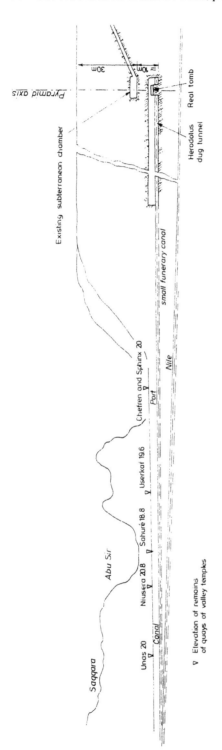

Figure 39. The 'horizon' of Cheops, with the elevation in metres of the valley temples: it was no accident that they were all, within a metre or so, on the same level over such a considerable distance. This proves the existence of a branch canal, in the light of which the words of Herodotus should be interpreted (Not to scale horizontally).

Figure 40. This boat (Fifth Dynasty), driven by the north wind, has just ascended the Nile against the current and is making its way towards Memphis. Before entering the canal it has furled its sails and lowered its mast.

enter the link canal: on both sides could be seen the still damp plots in which the peasants were treading the grain. To the north-west lay the ramparts of Memphis. Soon the boat would reach the horizon of the 'White Wall' nome and pass alongside the western ramparts of Memphis, where the gateways to the city were located. Large boats would tie up in the port of Memphis (Fig. 42) while others would continue their way northwards, past the small landing stages serving the temples in the valley, until they came to the port of Cheops.

This canal was a magnificent waterway with a host of vessels plying to and fro, from small barks made of papyrus to larger ones for livestock or pairs of boats lashed together for the transportation of gigantic blocks of rough-hewn granite from the quarries of Aswan. We can imagine the Pharaoh Cheops taking only a few hours to travel from his palace in Memphis to the port of his pyramid in a light skiff manned by the strongest rowers in the kingdom (Fig. 41). The pyramid texts refer to such trips: 'The King sails on his canal' (Pyr. 697) and 'You sail on your canal like Ra along the pathways of the firmament'.

Navigation on the Nile itself, on the other hand, was more of an adventure. When we think of all the activities of a great nation of boat-users, our judgment is warped by our experience of an industrial society that has forgotten all about the hard lives of boatmen and the difficulties faced by travellers over thousands of years. Let us listen for a moment to Madame de Sévigné in seventeenth century France who, to get to 'Rochers', took the boat at Saumur and described the trip in a letter to her daughter: 'It is twenty leagues from Saumur to Nantes and we resolved to do them in two days and to arrive in Nantes today; with this intention, we yesterday did two hours in the dark; we ran aground and remained at two hundred paces from our hostelry without being able to get to it.... We re-embarked at day-break and found ourselves so perfectly embedded in our sandbank that it was nearly an hour before we were able to resume our conversation; as we wish, come what may, to get to Nantes, we are all rowing.' Madame de Sévigné would have been amazed to learn with what ease, four thousand years before, the Pharaohs travelled along the canal from Memphis to Giza.

Figure 41. Rapid vessel of the kind probably used by the Old Kingdom Pharaohs.

Perunefer, port of Memphis and major port of Egypt

Did the lateral canal continue as far as the sea? Yes, it did, if we are to believe Herodotus: '*When the Nile overflows, the whole country is converted into a sea and the towns...alone remain above water... At these times water transport is used all over the country instead of merely along the course of the river, and anyone going from Naucratis to Memphis would pass right by the pyramids*'.

Why? The answer is that although certain settlements stood out above the inundation, other outcrops of higher ground were barely submerged and represented a great danger to navigation. Moreover, the current beside the pyramids would be less strong and the channel of the canal marked by beacons.

Another question: where would the Pharaoh's fleet be moored during the inundation? Historians have frequently confused ports and ornamental lakes:[22] the most favourable site for the construction of a port on the right bank appears to have been Coptos. On the left bank, however, the biggest port was that of Memphis. A bas-relief of the Sahouré causeway (Fifth Dynasty) depicts the return of the fleet to Memphis after an expedition to Lebanon. Although there is general agreement about the existence of a port, its exact location is in dispute: some scholars say it was in the plain, near but south-west of Memphis. I disagree.

In my view, it would have been impossible, in the eight months between two inundations, to finish building sea-going vessels, some of them over 60 m long, and it is hard to imagine the existence of large wharfs or docks along the river liable to be choked with silt at each floodtime. The Pharaohs needed to create a large naval base with a constant water level. This suggested the idea of excavating a wet dock jutting into the Libyan escarpment at right angles to Memphis (Fig. 42).

The Wadi al-Taflah, a minor tributary with steep sides lying due west of

---

22. Amenophis III ordered the building south of Thebes of the city of Malgata with its royal residence and a port 2.4 km long by 1 kilometre wide, but it was probably more for pleasure than for a naval base.

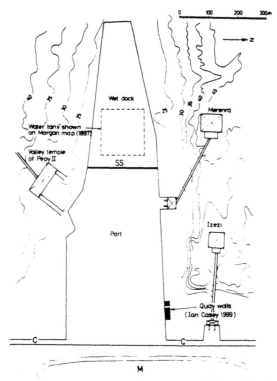

Figure 42. The site of Perunefer, important port of Egypt, was hidden behind the ramparts of Memphis. The rocks were extracted more particularly from the northern slopes of the wadi, which faced east. Another important function of the port was to serve the valley temples of two of the three pyramids shown. SS = wall submersible by floodwaters; C = canal; M = Memphis.

due west of Memphis, was the natural choice of location. In digging out the port, the stones extracted were used to construct the ramparts of the city. As for the idea of a wet dock, this was not revolutionary in any way but the outcome of observing the small dykes constructed by the peasants in the plain: after having been submerged they retained the water. In the case of this port, a wall was submerged by the rising floodwaters. The submersible step to be found in certain present-day marinas subject to tides is another version of the same idea.

It was an intelligent project, very carefully thought out and worthy of its name, Perunefer. In the Egyptian language, Nefer (*nfr*) meant beautiful (many beautiful Egyptian women merited the name Nefret or Nefer, among them Nefertiti and Nefertari).

This Perunefer, in accord with the topography of the site, also reflects the rationale of the Pharaoh's naval dockyards during the Sixth Dynasty. Pepy I decided to build a great capital city surrounded by ramparts: the wadi was cleared of its rocks, which were used to build the magnificent walls that aroused the admiration of the people. The capital city was called Mennefer, which means 'durable beauty'.

Later, two Pharaohs decided to build their pyramids near the new port, Merenra to the north, Pepy II to the south. The landing stages of their valley temples became essential features of the port, as we learn from history. Uni, the Governor of Upper Egypt under Merenra, had been sent on a mission by his king: 'Your Majesty sent me to Ibhat to bring back the coffin of the living, who is the lord of life, with its cover and the precious pyramidion for the pyramid.... Merenra appears in perfection in six wide boats, three barges, three vessels of 80 cubits in length... I cut down acacia to make a raft 60 cubits long and 30 wide, which was done in seventeen days during the third month of the summer. At the time when there was no water on the sandbanks I moored up against the pyramid' (Roccati 1982, pp. 196-197).

In 1998, Dr Ian Casey, Director of the South Saqqara Project, found, immediately south of the Djedkaré-Isesi causeway on the northern edge of the entrance to the Wadi al-Taflah (Fig. 42), lengths of a quay wall laid in sections with horizontal coursing.[23]

The testimony of Piankhi, who reigned from 751 to 730 BC at the end of the Third Intermediary Period, confirms the siting of the port behind the capital.[24] We have already spoken of this Pharaoh, who came from the north to conquer Lower Egypt: amazed by the strength of the city's walls, he had gone against the counsel of his advisers and taken the right decision to attack the city from behind, from the west, 'where the port is situated'. There he found the fleet moored.

Deep in this creek, boats of all sizes – canoes, fishing boats, cargo boats, troopships up to the longest sea-going vessels – were constructed in a wet dock throughout the year. It should not therefore come as a surprise that Memphis is described in the texts as a favoured point of departure for naval expeditions. Here, too, ships that catch the imagination were constructed: 'I am constructing new vessels for my voyages and I am covering them with gold so that they will illuminate the Nile'.

The site of this port is today occupied by a large palm grove. In the shadow of the trees one is tempted to daydream about the fantastic city of Memphis and its port, which embodied the nautical ambitions of the Two Lands, for here lived a people of mariners. Thousands of years ago, the air reverberated with the cries of these sailors, with the blows of axes, the squealing of copper saws and the thudding of mallets.

The Egyptian forest does not abound in timber of good quality. As we learn from Uni, they had to make do with acacia, a hardwood that is difficult to bend into shape or cut into long planks. The planks had to be short and, according to Herodotus, 'The Nile boats are built of acacia wood...[from which] they cut short planks, of about three feet long,... and lay them together like bricks'.

---

23. Bulletin of Egyptian Archaeology 1999, p.24.
24. See J.H. Breasted 1906, IV, pp. 432-5, which contains a translation of the Victory stela.

Figure 43. In the left-hand panel, a tree-trunk is being roughly shaped; in the centre panel, rough patches on the hull are being smoothed out; on the right, the bulwarks are being fitted.
(Photograph J-F. Livet).

Figure 44. The 'bricks' of Herodotus: construction of a boat with short lengths of wood.

This Meccano of planks pegged together[25] formed a hull without keel or ribs (Fig. 44).

The voyage to the land of Punt to bring back the precious myrrh was one of the main preoccupations of the kings. To get to the Horn of Africa and Yemen required a naval expedition. For this purpose, large vessels were transported in

25. The planks were probably not held together by thin cords threaded through holes to midthickness, as is suggested by the reconstitution of the Cheops funerary barge. In that case it appears to be a kind of symbolic wickerwork not typical of the naval architecture of sea-going vessels.

pieces from Perunefer and the arsenals in Coptos to Berenice on the shores of the Red Sea across a hostile desert valley – the Wadi Hammamet. The stela of Henu, who travelled under the orders of Montjuhotep III (Eleventh Dynasty), bears the following inscription: 'My master sent me to launch sea-going vessels for a voyage to Punt to bring back fresh myrrh for him'. Henu reached the Red Sea: 'Then I reached the sea and put together the sea-going vessel and dispatched it laden with all manner of things'.

But were not the hulls of these ships in danger of being broken by the waves of the open sea or by the sandbanks of the Nile? To forestall these dangers the ingenious Egyptians used a technique that today we would call prestressing: powerful ropes running from the prow to the poop were tightened by twisting to make the ship as taut as a bow. The system of ropes rested on a series of stanchions so that, when the sea was rough, the ropes could be tightened by turning a spar slotted through the separate strands. This system of adjustable prestressing was several thousand years ahead of its time.

WATER AND FUNERARY RITES: THE TOMB OF RAMESSES II AND THE HORIZON OF CHEOPS

The Egyptians spent more time in preparing their eternal resting-places than in arranging their homes. Diodorus (Book I, 2) tells us that 'The inhabitants of the country attach no importance to the time they spend in this life but are greatly concerned about the time that, after their death, will perpetuate the memory of their meritorious deeds. The houses of the living they call 'inns' because we do not inhabit them for very long. But the tombs of the dead they call eternal abodes. That is why they set little store by the appointments of their homes but neglect nothing for their tombs'. Prince Hardjedef gave the following instructions: '*Make yourself a fine dwelling-place in the cemetery, make yourself a worthy place to the west, for the house of death is for eternal life*'.

Pharaoh after Pharaoh sought to surpass his predecessor in the architecture of the tombs. This is easy to see in the Valley of the Kings (increase in the number of rooms and in the number of pillars to hold up the heavens, a growing richness of decoration) and, were more evidence needed, the recent discovery of a hundred and fifty burial chambers for the direct descendants of the prolific Ramesses II are further proof of the extraordinary dimension of certain funerary structures.

Why this urge to excel? The celebrated expression used for those setting out on the great journey is well known: 'You did not leave as a dead person, it is alive that you set out'. The Pharaoh, lying prone, well swaddled, will spend centuries in the dark after only a decade or two as king. But he is confident of his future life. Chapter 110 of the Book of the Dead, also called The Funeral Ritual, describes the other world in which humans would continue to live. It is the place where *a person may be powerful and happy, may work, harvest, eat, drink, make love, and do everything that is done on earth*. To reach the land of the blessed, the dead had to cross a canal, and if the god Osiris considered them 'just', they were allowed into the 'Field of reeds'. A text of the Twenty-second Dynasty confirms that the dead Pharaoh may take with him one servant for each day of the year, to be guarded by thirty-six overseers: he dies peacefully, in the certainty of entering an eden where he will be spared all weariness.

After the Pharaoh's death, his body was placed in a large royal barge which, followed by a large fleet, sailed towards the holy sites; after further purification, the body was handed over to the priests to be borne to its final earthly resting place.

The tomb of Ramesses II

Ramesses II had chosen for his tomb the lowest site in the Valley of the Kings, for he wished in death to be the *first among equals*, at the very entrance to the Valley of the kings. In contrast, we may note (Fig. 45) the discreet site at the top of the valley chosen by his predecessor Thutmose I, valiant warrior and the first Pharaoh to choose that valley for his eternal home. Ramesses' desire to be pre-eminent would prove fatal.

The forces of evil dwelt in the chosen site: or, more precisely, the geological and hydraulogical factors in that particular place were most unfavourable.[26] Whereas all the other tombs in the Valley are cut into the mass of the Theban limestone, a first-rate stone, in this case, because of the low elevation of the site, the floor of the large chamber lies at the level of the underlying Esna shale. When this shale is imbibed with water, it swells and exerts a very considerable pressure: this is precisely what occurred, for rainwater seeped down through the ceiling of porous limestone, with the result that the tomb of Ramesses II has become over the last twenty years or so a classic example of errors to avoid in applied geology (see Fig. 48). The shale on which the pillars were founded reacted to the humidity by exerting enormous pressure, forcing the pillars upwards into the vault. Conversely, during the dry season, the underlying shale retracted and, in descending, pulled at the pillars: a *danse macabre* within the tomb of the great Ramesses II.

The burial chamber was a noble design: a large room with a basket-handle arch ceiling supported by eight pillars, with side chapels that foreshadow the ambulatories of our churches. But such slender supports could not withstand the pressure of the vault. Whereas the burial chamber in most tombs has only six pillars, this one has eight, as if Ramesses had a premonition that he was risking a dangerous challenge to the laws of soil mechanics.

After having been looted – the supreme insult – the tomb was almost entirely filled with mud brought in by occasional torrential rains: detritus and gravel invaded the tomb, which became so to speak the sewer of the Valley.

---

26. A detailed study of the process by which the chamber gradually collapsed was presented by G. Curtis and J. Rutherford to the Tenth Congress of Soil Mechanics and Foundation Engineering in 1981.

Thutmose

Ramesses II

Sons of Ramesses II

Figure 45. Location of the tombs of Thutmose I, Ramesses II and the latter's sons in the Valley of the Kings.

It was said that Ramesses II would be spared nothing: about a century after his death, in the twenty-fifth year of the reign of Ramesses IX, his mummy, in danger of being looted, was taken out of his sarcophagus and subsequently transported by the High Priest Pinedjem, in about 970 BC, to the celebrated hiding place at Deir el-Bahri, which remained inviolate until the nineteenth century. A third voyage then took him to the Museum of Cairo, a fourth to Paris and a fifth and last brought him back to Cairo. Poor wandering soul!

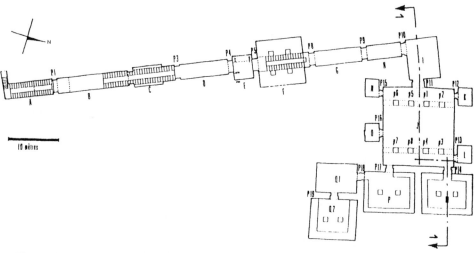

Figure 46. Plan of the tomb of Ramesses II, according to the measurements taken by the Theban Mapping Project.

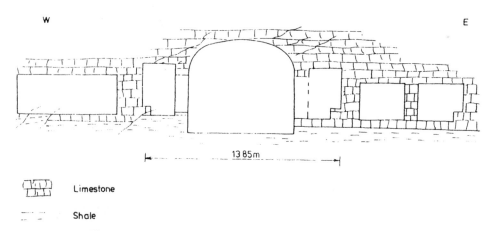

Figure 47. Vertical section along the west-east axis 1-1 in Figure 46.

A small team of researchers from the French National Centre for Scientific Research (C. Leblanc and A. Guillaume), in association with the Louvre Museum and initially with the backing of the Elf Foundation, has undertaken to clear and restore the tomb of Ramesses II. It is intended to install a museum in it to satisfy the curiosity of the living: under the restored vault it will be possible this time to breathe in the scent of immortality and reflect on the ambitions and errors of the great king.

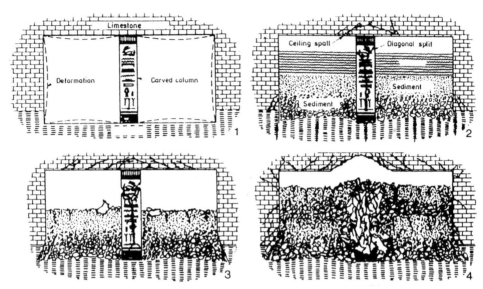

Figure 48. Vanitas vanitatum! The successive tortures inflicted on the burial chamber of Ramesses᾽ II by natural forces (Curtis and Rutherford, Brooklyn Museum, 1981). Phase 1. *Initial excavation.* The roof of the chamber has to bear the weight of over 280m of rock, and this has resulted in minor distortion and the concentration of forces at the corners. Phase 2. *Flooding and swelling.* The saturated shale swells and exerts enormous pressure on the pillars and partitions, splitting walls and columns. Phase 3. *Drying out.* The drying shale slowly contracts, leaving portions of the pillars broken and walls hanging from the roof structures. Phase 4. *Collapse.* After several cycles of flooding and drying, the ceiling crumbles and the hanging members drop into the flood debris...

The Pharaoh had fifty two sons and the tomb built for them (Fig. 45) on the other side of the entrance to the valley is of considerable size, with 110 small chambers off an immense hall of 325 m^2 supported on sixteen pillars. Because this immense hypogeum is located at a higher level, it has had a happier fate.

The burial chamber of the Pharaoh Cheops

Herodotus (Book II, 124) speaks of the existence of 'underground sepulchral chambers in the hill on which the pyramids stand: the king made these chambers for his tomb and, in order that they be on an island, he brought water from the Nile by means of a channel (αυλωνοσ εσω)'.[27]

No Egyptologist has believed in this affirmation. Some think that Herodotus misinterpreted a myth according to which the Pharaoh, after his death, sojourned

---

27. Adapted from the French translations of Herodotus since the English is misleading in this passage.

on an island called the 'Island of Enchantment'. It is rightly argued that the chambers in question must have been excavated above the water table but, since the level of the water table lies above that of the Nile flowing down the valley opposite, how could the water from the river be made to reach the chamber? This is true, so we need to take a closer look at what the Greek historian is saying.

Poor Herodotus! All those who sing his praises as 'the father of history' seem to derive a malicious pleasure from pointing out his mistakes, his credulity and even his bias. In actual fact, what he has to say is in most cases true or at least explicable; as Jacqueline de Romilly puts it so well, 'Peu de science en écarte, davantage de science y ramène' (when your knowledge is slight you reject him but when you know more you bring him back). In this particular case, he has misled future exegetists with his concision. To fully grasp what he is talking about we need to understand the topography of the valley and the logistics of the worksite, for which the canal we have already referred to was crucial: the canal reach of Cheops, being horizontal, was on the same level as the Nile at the point where the canal left the river, *much further upstream*. When this reach of the canal arrived at the port of Cheops its level was well above the plain and the water table, which descended towards the Nile, underneath the hill, with a very slight slope. There was therefore nothing to prevent the chamber mentioned by Herodotus being constructed in the dry. An extension to the canal started from the port and turned slowly to form a 'dog channel' leading to the underground burial chamber. This channel was blocked up before the death of the Pharaoh. On the floor of one of the chambers would have been left a raised platform of solid rock that, on the death of the Pharaoh, would have been surrounded by water from the canal. On this water would float the funeral barge bearing the sarcophagus. All this explains the testimony of Herodotus, which is supported by a pyramid text: 'The underground resting-place was dug out for the King. Make a passageway for the King'.[28]

This solution was both ingenious and in perfect accord with the position of his pyramid which, unlike the others, lies at the edge of the escarpment in order to reduce the length of the tunnel or 'dug channel'.[29] The solidity of the rock made it possible to excavate the chambers without danger of their collapse under the weight of the pyramid. Cheops was a better geologist than his father Seneferu and his successors Amenemhat III[30] and Ramesses II.

---

28. Pyramid text 2251.
29. The distance from the north-east ridge of the pyramid to the edge of the escarpment is barely 100 m and it was precisely there that Howard Vyse, who showed great interest in the account left by Herodotus, tried in 1837 to find the 'dug channel'; but the great volume of rubble that Chephren had had tipped there, perhaps with the purpose of discouraging looters, led him to give up.
30. It will be remembered that the tunnels under the pyramid of the Pharaoh Amenemhat III were seriously damaged by the weight of the structure above: the ground at that spot had the same weaknesses as at Dahshur South, built on a nearby site, whose burial chambers were badly damaged.

The horizon of Cheops

The expression *Akhet-Khufu*, found frequently in the texts, is sometimes translated 'horizon of Cheops' and sometimes 'tomb of Cheops'. The analysis proposed here gives the expression its full significance: the horizon of Cheops is the
level of the water in the canal that ran from Memphis to the Pharaoh's burial
chamber. This also explains why the translators sometimes took the word as referring to the tomb.

The importance of what Herodotus says

Many details given by Herodotus after his numerous voyages, which not so long
ago seemed highly implausible, are now, after further research, acknowledged as
being true.[31] During his travels to Persia he had a translation made of the royal
cartouche of Xerxes which was subsequently crucial to the deciphering of the cu-
neiform script. As Jacques Lacarrière rightly remarks,[32] the illustrious Greek did
not come to Egypt, like many other Greeks before him, to open a trading post but
for the pleasure of informing us and enriching our knowledge, and the journey required considerable courage.

In 1837, the Englishman Howard Vyse, discoverer of the chambers above the
King's Chamber in the Great Pyramid and convinced of the truth of Herodotus'
account, had a shaft sunk from the existing underground chamber, but the heat
and lack of air were unbearable and the operation had to be quite quickly halted.
Even if he had continued he would have had little chance of piercing the roof of
the real burial chamber because his shaft was out of line with the pyramid's north-
south plane of symmetry. That axis always seems to be reserved for something
important. Indeed, the corridors and the Grand Gallery appear to have been sys-
tematically displaced, as if the Pharaoh wished to keep the central location for his
tomb, with the island and sarcophagus placed directly under the apex of the
pyramid where the two planes of symmetry meet: for an Egyptian of that time, the
Pharaoh reigned over the whole world and the axis of his pyramid symbolized the
axis of the world.

With modern methods of temperature control and ventilation, it would not be
difficult to sink a shaft in the vertical axis of the pyramid from the existing underground chamber (Figs 39 and 49); it could be done quickly and relatively cheaply.
Indeed, this is the only feasible way of proceeding since it would be a hopeless
task to look for the opening of the 'dug channel' in the hillside owing to the

---

31. Examples include the dimensions on the ground of the Tower of Babel, the Royal cartouche of Xerxes
    and a number of other details mentioned by Jacqueline de Romilly ('Introduction aux oeuvres
    d'Hérodote et de Thucydide', la Pléiade, 1964, p. 28) as well as the pronunciation of the name of the
    builder of the Great Pyramid in which no one believed until the discovery in the nineteenth century,
    inside the pyramid, of the cartouche of the Pharaoh.
32. *L'Egypte. Au pays d'Hérodote*, Ramsay 1997, p. 15.

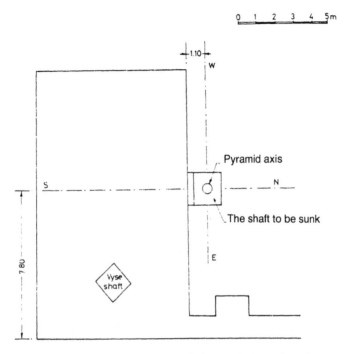

Figure 49. Soon the truth of the words of Herodotus will be revealed and the existence, at an elevation of about 20 metres, of the parallel canal with its funerary extension demonstrated (see Fig. 39).
Then will it be felt necessary to sink a shaft under the apex of the pyramid from close to the north
wall of the existing underground chamber, whose floor lies at an elevation of about 32 metres. After
some eight to ten metres the shaft will pierce the ceiling of the real tomb of Cheops, revealing one
of the new century's greatest discoveries.

enormous quantities of rubble that have been pushed over the edge of the plateau
by bulldozers; most of this rubble is made up of lumps eroded from the pyramid
by the wind and the sun. It is as if the pyramid wanted to help safeguard its own
secrets.

Before leaving these two famous Pharaohs to the mysteries of their eternal
resting-places, it is possible to draw a few conclusions.

Ramesses II, confident of the tribute that would be paid to him by posterity,
chose for himself a tomb that, he thought, would be treated with respect until the
end of time; an illusion at the close of a very long reign. The grandeur of the work
led to an error of judgment. The tomb had two weaknesses. Firstly, the ease of access tempted looters and, secondly, the choice of a hostile geological setting made
its eventual destruction inevitable.

On the other hand, the tomb of his distant predecessor Cheops, as described by
Herodotus, has a powerful symbolic significance in keeping with the thought of
the Amduat, the underworld, literally 'the hidden chamber', to which the Pharaoh
travelled across the watery regions. Here water, the peaceful water on which the

Figure 50. The steep edge of the plateau of Giza in about 1857 (Photograph by Robertson and Beats from the collection of the Banque du Louvre, Paris).

Figure 51. The mound of rubble that covers the slope today.

funeral barge floated, is the element of resurrection. In contrast, in the tomb of Ramesses II, water was the element of destruction. At Abydos, where various inscriptions show Ramesses II's pious respect for his father Seti I, the latter's cenotaph lies on an island surrounded by a canal. This is the Osireion symbolising the primeval mound rising from the waters at the creation of the world. Cheops knew this place of piety because it was there that the only known statuette of the Pharaoh was found.

THE PHARAOHS: EXAMPLES OF THEIR WISDOM AND SKILL

There were thirty dynasties and over one hundred Pharaohs between Narmer (3000 BC) and Nectanebo II (360 BC), the last king of Egypt before the second period of Persian domination – over twenty centuries from which we have to subtract three Intermediate Periods (2180-2060, 1785-1580 and 1085-666 BC) about which relatively little is known. For about twenty centuries, then, over a hundred kings succeeded one another. They were not necessarily related, as some had no heirs and some came to the throne as the result of a coup d'état or an assassination of which we know nothing. Owing to the great genetic diversity of the Egyptians it is not surprising that we find Pharaohs as different as, for example, Thutmose III, the great strategist who led his troops as far as the banks of the Euphrates, the ostentatious Amenophis III, the mystical Akhenaten or the general Horemheb.

We still know nothing about the mixture of beliefs and calculations that guided the action of the Pharaohs or about how they enjoyed themselves in life. Ancient Egypt is like a mirror that reflects the image of one Pharaoh and then another. Each of them stands for a particular period. For many people the Pharaoh represents greatness, beauty, victory; for others, he is a horrible tyrant, the persecutor who drove out Moses; feminists will remember only the image of Queen Hatshepsut or Cleopatra. As for Herodotus, he liked most of all people to speak to him about the 'cursed kings' such as Cheops. Nor should we forget the Pharaoh of the golden age, Amenophis III, when Egypt amassed the tributes of gold imposed on the Nubians and he became the richest man in the world: 'I cast statues made of gold and electrum decorated with lapis lazuli and precious stones'. All this gold fostered illusions: when he celebrated his many jubilees he commanded portraits of himself that made him look younger and younger.

Others will remember only the hubris of Cheops, which they will compare to that of Ramesses II: a risky comparison for they were as separated in time as Napoleon and Charlemagne. And yet they were two Pharaohs who were enamoured of grandeur both in this life and in the after-life; they were majestic in the extreme but the nature of their hubris was different. Again and again Ramesses sought to exceed all bounds: he was the greatest of the builder-Pharaohs, he had a very long reign but in the end he tends to stifle our imagination – his statues are heavy and massive, as if his artists had been ordered to limit themselves to the colossal. So much so that today, when someone excavates an enormous statue without a name on it, it is immediately attributed to Ramesses II. The hubris of Cheops is not to be measured by the size of his pyramid alone; that construction was only one aspect of his immense genius. Each person will have a Pharaoh he admires or prefers, for the world they represent is so diverse and mysterious. It is that very diversity and mystery that explains our undying admiration for this ancient civilization.

Our attitude to Egypt is subjective and reflects our reading and our travels. Even today, despite the great contributions of Egyptologists from every country, our vision of ancient Egypt remains coloured with this subjectivity, reflecting the special fields and the nationalities of the foremost researchers.

Leaving aside these 'star' Pharaohs, these sun kings of Egypt, and their contrasting qualities, let us see if it is possible to discern a kind of Pharaonic wisdom that was common to all of them.

AUTOCRACY, PATRIOTISM AND DEVOTION TO MA'AT

The Pharaoh is an autocrat who rules the world: he is the owner of the soil, of his subjects and of the animals; he is identified with the gods and takes on their virtues – he is not only Halac-Uinic, 'the one who knows' of the Mayas, but all his knowledge come to him from the gods. As Hegel wrote, 'It is in Egypt that, for the first time, a kingdom of the invisible was established'. In Babylon, the Tower of Babel had at its summit a large temple where, according to Herodotus, there was no divinity; in contrast, the pyramid had a religious and political purpose and symbolised the Pharaoh communicating with the heavens and reaping celestial powers.

Despite all these myths, the Pharaoh remained a mortal being forced to delegate his authority; but many ordinary Egyptians, the victims of shocking injustices, would eventually be smitten with doubt.[33]

In the history of ancient Egypt, as Bernadette Menu[34] remarks, the good Seneferu was at first contrasted with the nasty Cheops. Was it at this point that everything began to go wrong? The papyrus containing the story of Neferirkaré's love affair with his marshall shows that a Pharaoh could sometimes escape from his Olympus. Nor were the works of the Pharaohs always faultless: in the Great Pyramid, for example, a disaster in the King's Chamber cast doubt on the infallibility of the Pharaoh Cheops[35] with the result that, after his death, the people took their revenge by shattering the statues of him and the priests insisted that his successor should no longer be 'Ra', the Sun, but only the 'Son of Ra'. This amputation

---

33. We find this sentence in the *Dialogue of a madman:* 'Death is today before me like the healing of an illness, like the first time out after suffering'.
34. Cf. 'Individu et pouvoir en Egypte pharaonique', *Méditerranées*, 13, 1997.
35. J. Kerisel, *Génie et démesure d'un pharaon: Khéops*, Stock, 1997.

of a title was more significant than is thought because the sun represents the energy and strength displayed by all the gods. Perhaps it foreshadowed the chaos of the First Intermediate Period from which Egypt recovered to find itself with a less despotic Pharaoh and a more powerful priesthood. Ma'at would continue to embody the standard to attain and the idea of moderation at all levels of society.

'I made the great inundation so that the humble might benefit by it like the great'.[36] The Pharaoh was the prince of water, 'he who gives water to the soil'[37] and, with the coming of the inundation, he will set in motion the process of flooding and drying that will give life and prosperity to all the people. This explains why the hieroglyph that represents Ma'at is a basket.

The First Intermediate Period is known to us from the existence of a few texts, one of which is the *Admonitions of Ipuwer:* 'It is a time of misery. All is destroyed, the social order has been overturned, servants take the place of masters, the public buildings are burnt down, the pyramids are looted'. It was a Pharaoh originating from the south who would take up the torch. He took the name of Amenemhat I. He understood that, in the palace, Isfet (injustice) must give way to Ma'at (order, law) and he would inspire a series of surprising texts, known as the 'Instructions': real social justice must be based on the benevolence and solicitude of the aristocracy towards the lower classes of the society. We also find a hymn to productivity that is still relevant: *'There is not a single ploughed field that ploughs*

Figure 52. Liberation from the yoke of the Hyksos. Ahmose expelled them from the country after an occupation lasting about a hundred years.

---

36. M. Lichteim, *Ancient Egypt, Litterature 1*, 1973, p. 132.
37. A.H. Gardiner, *Admonitions of an Egyptian Sage*, p. 55.

itself... What exists is created by human beings. Our lives depend on what is in their hands...' We may take it for granted that the ideas of this liberal Pharaoh aroused strong criticism because he was later assassinated. His immediate successors, however, would continue his work until the Middle Kingdom collapsed into a Second Intermediate Period.

The country did not lack liberating Pharaohs. After the period of troubles, several figures stand out: Montjuhotep and Amenemhat I after the First Intermediate Period, Ahmose and Thutmose III after the Second. On the other hand, we find a Queen Ankhes-en-Amun, daughter of Akhenaten and widow of Tutankhamen, appealing to the king of the Hittites to send one of his sons to marry her. In spite of this exception, autocracy associated with patriotism and justice was characteristic of the Pharaohs.

Respect for the past

Under the firm rule of the Pharaohs, Egypt always paid a lot of attention to its past, which was the subject of numerous archives. The country honoured ancestral values: 'Live like your fathers and ancestors, may you work... in accordance with the tradition. Open the books to read them and respect the tradition.[38] May you be a scribe and frequent the House of Life; may you become like a chest of books'.[39] This memory of the past so astounded the Greeks that Plato put the following remarks into the mouth of an Egyptian priest: '*You Greeks are like children for you have no ancient history, whereas here nothing great or beautiful was ever accomplished without the memory of it being preserved in our temples*'. The Egyptians and their Pharaohs tirelessly repeated: 'Our ancestors have preceded us'.

A rational desire to excel

Frontiers

The Pharaohs were constantly exploring beyond the frontiers of the country – southwards up the difficult cataracts to attain Nubia, towards the north-west and north-east into Libya and Sinai. The natural frontiers were solemnly marked by stelae and future Pharaohs were firmly enjoined to guard them carefully: '*He who allows himself to be driven out of his frontiers is a true coward... Any son of Us who preserves this frontier established by Our Majesty is truly Our own son engendered by Our Majesty*'.

In addition to being the protector of new frontiers, the Pharaoh became he who enlarged his inheritance and sought to extend the frontiers of his predecessors but also to surpass them in art and architecture. Thutmose I of the Eighteenth Dynasty

---

38. Ph. Derchain, *Revue d'Egyptologie*, Vol. 40, 1990, p. 44.
39. P. Posener, *Revue d'Egyptologie*, Vol. 10, 1955, pp. 64 and 68.

stated: 'I am doing more than all the other kings who came before', and the young Tutankhamen affirmed that he 'has surpassed what has been done since the time of his ancestors'. This claim of the short-lived Pharaoh contains an element of boasting typical of the autobiographies.

Architecture

The size of the pyramids gradually declined from the Fourth Dynasty onwards but now they were coupled with other major works: the walls of Memphis, temples and palaces and other large public projects. At first the temples were located in caves or built of brick before becoming buildings with a longitudinal axis that were continually enlarged and never finished. Although four masts in front of the entrance pylon were generally considered to be enough, Akhenaten erected ten in front of the temple dedicated to his single great Aten, the Sun-Disk.

The temples were made bigger: the texts say that they attained the stars and touched the sky. But even with this increasing size, the main lines of the architecture remained the same. We find the two most important aesthetic principles which the Latins called *firmitas* (solidity in conception and construction) and *venustas* (beauty on account of the harmonious proportions of its parts) and, under the Egyptian sky, a third quality, *suavitas colori*, beauty of the colour reflected on the stone.

But for both pyramids and temples there was above all the 'revealed' and the 'hidden'. The revealed is what the eye sees – in the case of the pyramid, the crystal polyhedron shimmering in the light. But that was usually just the surface architecture, a handsome skin that masked something less noble: the fine limestone casing was no more than a disguise to hide the most ordinary and sometimes very careless[40] masonry; in contrast, the temple housed the divinity, the creative force; in entering the 'naos' the Pharaoh passed through 'the gateways of the sky' in order to ask the gods to give Egypt the vital energy it needed. The temple would soon become the abode of Ammon, 'the hidden one', whose true nature is inaccessible to human beings.

Art

On the subject of Egyptian art, everything or almost everything has been said: its quality is as apparent in the tiny hieroglyphic pictographs as in the large reliefs of

---

40. This fact of Egyptian architecture continues to mislead many specialists with their scale models, who imagine the pyramids as superposed parallelepipeds shaped like sugar loaves or as an immense set of building blocks. The trick of hiding poor quality masonry behind a magnificent facade has been by no means uncommon in history: it was used by the architects who built the towers of the Italian republics in the twelfth century, the Tower of Pisa in particular, the great masters of the Renaissance, and even Jacques Soufflot (1709-1780) for the rubble-filled pillars supporting the dome of the Pantheon in Paris. They all had recourse, not without accidents, to this dual-quality architecture in order to cut costs, but at the expense of security.

the Egyptian tombs, and it achieved a magnificent elegance at the time of Amenophis III and his son Akhenaten. It only needs to be added that it was not an art of the easy solution: statues were sculpted in the most thankless materials, that of the chancellor Nakhti (about 2000 BC) in a trunk of acacia, for instance. Artists even pitted themselves against the hard granite of Aswan or even against diorite, the hardest stone of all, so hard that it was used to polish all the other stones.

The Pharaohs liked to be portrayed in such rocks born in the fiery depths of the earth. Large statues expressed their power. But as soon as the artist is given more freedom, he is capable of choosing the most suitable material for his subject. The finest examples are often in limestone, a stone resulting from the accumulation of billions of once living creatures: it is a living rock par excellence and with it the Egyptian artists gave life to their subjects. But they also used sandstone, with its fine grain capable of expressing sensitive features. In the Louvre Museum one can admire from close quarters 'the kneeling scribe' and a portrait of Akhenaten, the former in painted limestone dating from 2600-2350 BC and the latter in painted sandstone dating from 1350 BC, two masterpieces remarkably well displayed. Around the summer solstice, the sun in Paris sets behind the Arc de Triomphe.

Figure 53. Akhenaten and his family under the rays of the sun.

Figure 54. Akhenaten.

But before Ra rejoins the world of shadows, he lights up the Sully wing of the Museum and brings these two figures to life. The eyes of the scribe then gleam with an intense look[41] while Akhaneten is pale and dreamy until his face becomes pink when it is lit by the last rays of the sun he loved so deeply.

Technical research

We have already shown how advanced the Egyptians were in various fields, especially in everything related to water. Here we would like to draw attention to a particular domain in which their ingeniousness was tinged with symbolism.

---

41. In the white limestone of the cornea is encrusted a cone of crystal hollowed out to a point inside: using their knowledge of the anatomy of the eye, the Egyptians succeeded in accurately reproducing the curve of the cornea and the diameter of the pupil.

The system of counterweights used by Cheops: The purpose of the Grand Gallery
Transport by water did not solve all the problems: the materials unloaded at the
quayside in the port still had to be conveyed to the religious communities surrounding the pyramids along lengthy and steep causeways.[42] The loads were
placed on sledges which were pulled over the ground by teams of hauliers. Whole

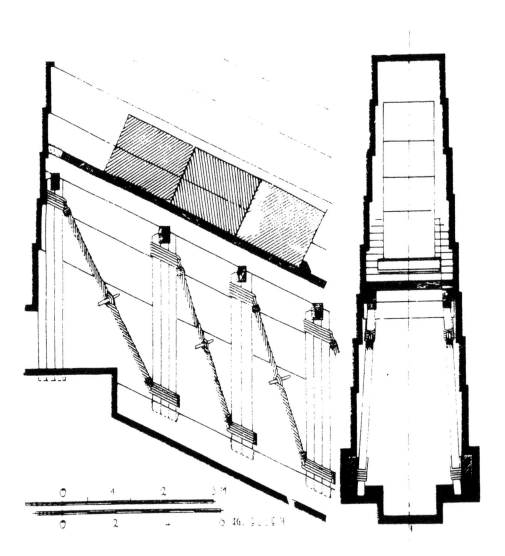

Figure 55. Three stone plugs sliding on a wooden floor supported on props, as drawn by the Egyptologist Ludwig Borchardt in 1932.

---

42. The word 'steep' needs a little explanation. The slope of the causeway ranged between 5% and 9%.
   Ergonometric experiments have shown that when the slope is over 15% most of the energy available
   to a human being is used up by the elevation of his own weight.

lives were devoted to the battle against friction between sledge and soil. This problem led the Egyptians to look for the best means of reducing the forces that so strained their backs.

In Egyptology, little attention has been paid to this problem. And yet it was one of vital importance in everyday life as well as in construction projects. It has often been claimed, after a few cursory experiments, that the sledges slid over the moistened silt almost as easily as on ice. This is not true, as we have already shown[43] elsewhere.

The construction of the Great Pyramid raised a truly exceptional problem – the need to haul up about sixty enormous stone slabs to be positioned above the King's Chamber. As they could not be hauled up by a system of levers made of wood, they had to be dragged up ramps on sledges. But each slab, which weighed around sixty tonnes, required a team of several hundred hauliers whose efforts were difficult to coordinate.

One of my readers, Mr Pierre Blondin, a civil engineer who used to have his own business, suggested that the raising of these slabs could have been coupled with the simultaneous lowering of loads towards the Grand Gallery, just as a lift or funicular railway will rise more easily when worked in conjunction with a descending counterweight.

It is true that the presence of the Grand Gallery within the Great Pyramid has never been explained to everyone's entire satisfaction. At the beginning of the twentieth century, the German archaeologist Ludwig Borchhardt devoted a long study to this exceptional and unexpected construction. In his view, the notches sculpted in the base of the Grand Gallery served to maintain props to support a surface with the same slope as the Grand Gallery; he thought that this was used to stock the huge stones that would later serve to mask the entrance to the ascending corridor so that, after the death of the Pharaoh, the procession would be able to attain the King's Chamber under this wooden floor.

The celebrated archaeologist is right, but he did not take his idea far enough. The Pharaoh had previously used this wooden floor for another purpose connected with the construction of his unique Chamber: along this floor, from top to bottom, the large stone plugs slid slowly down, attached by ropes to the sledge carrying the stone slabs up the construction ramp of the pyramid (Fig. 56). As the plugs descended they pulled up the huge stone slabs quite effectively owing to the steep incline (one in two) of the lubricated flooring, six times as steep as Rue St Jacques in Paris and four times as steep as Rue Lepic in Montmartre. The idea of letting heavy weights slide down a steep slope was not unknown to the Egyptians, as we can see from the relief (Fig. 57) depicting a sarcophagus being lowered towards the tomb before the watchful eyes of a divinity. The slope of the descending corridor under the pyramid of Abu Roash (Fig. 58), which is the same as that of the Grand Gallery, helps us to understand the effectiveness of the system.

---

43. J. Kerisel, *op. cit.*, pp. 200-216.

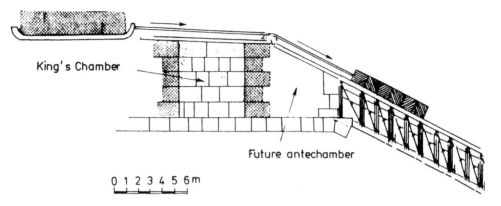

King's Chamber

Future antechamber

0 1 2 3 4 5 6 m

Figure 56. Hauling up the huge stone slabs used to cover the King's Chamber by means of counter-weights in the form of six stone plugs sliding down a lubricated wooden floor. The Chamber would have been temporarily refilled.

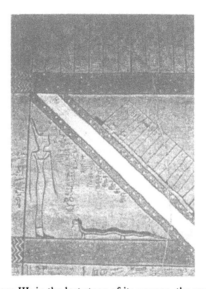

Figure 57. Tomb of Tuthmose III: in the last stage of its voyage, the sarcophagus descends towards its final resting-place under the protection of a divinity.

On a slope as steep as this, lubrication with a muddy paste had to be ruled out because it would 'creep' too much. This led me to undertake a whole series of experiments to test the friction of stone on acacia planks lubricated with animal fat, more particularly mutton fat: in the first place, a film of slightly warmed animal fat counters a large part of the resistance to initial movement that is perfectly normal in questions of friction and lubrication. Moreover, the friction does not increase but remains some five to six times less than with moistened clay. The re-

Figure 58. Under the pyramid of Abu Roash the descending corridor towards the burial chamber has the same incline as the Grand Gallery. (Excavations of Professor Michel Vallogia).

sults of these experiments are backed up by the experience of naval architects in the launching of large ships on slipways, still constructed along these lines and lubricated with synthetic grease. The Egyptians probably used the fat of cows: the goddess Hathor, who symbolizes fertility, is portrayed in the form of a cow. This extraordinary animal deserved to be deified for, after having spent its entire life pulling a swing-plough to till the soil, it then offered its fat to facilitate the efforts of those it had served and would later feed.

Did there exist ropes long enough to use this system of counterweights? The answer is yes. The Egyptians spent a lot of time weaving strands of papyrus to make ropes of all lengths. Evidence for this can be found in Figure 59, in which time is symbolized by an endless rope.

The counterweight would of course be composed of several blocks of stone attached together which could subsequently be hauled back up one by one (see Appendix 3).

The suggestion offered by my reader appears to be sound: the Grand Gallery is a work of art through its exceptional architecture but it is also extremely inventive in its function. It gives us a glimpse into the minds of the Pharaoh and his architect. Glory was very important to the Pharaoh and this required that the chamber

Figure 59. Time represented as an endless rope. (Book of Gates). From: A. Piankoff and N. Rambova, The Tomb of Ramesses VI, New York, 1954, p. 174, Fig.47.

inside the pyramid where the cenotaph would be laid should be large in size and covered by a fivefold roof of enormous stone slabs. It would be at this point that the Grand Gallery was conceived both as an aid for the construction of the chamber and as a symbol in which the daily heaving up of the blocks of stone served as a reminder that our world, dominated by chaos and the forces of evil, would slide towards nothingness if it were not recreated each day by the gods.

Sapiens means the one who knows but also, in ancient Egypt, the one who creates. Are we ourselves still worthy of our appellation *homo sapiens sapiens*? Let us admire these very ancient Egyptians who were *sapiens* in the full sense, in contrast to civilizations like that of Byzantium which lasted a thousand years but made no effort of research. That could not be said of ancient Egypt.

We shall see later that during the twentieth century there was a tendency to admire the exceptional scale of pharaonic architecture and to draw inspiration from it for the conception of dubious projects, forgetting that the great kings of Egypt respected the value of superior learning and wisdom.

THE DECLINE AND ITS CAUSES

Under Ramesses II,[44] the *pax egyptiaca* gave tribes coming from elsewhere security; the revolt in the land of Canaan five years after the death of the Pharaoh shows how much his regime had deteriorated with time and suggests that the old king had bequeathed a disturbing situation to his son.

At the time Egypt was being affected by a gradual change in climate, fauna and flora. The floods were becoming insufficient: compared to the price of metal, the price of wheat at the end of the Ramesside period was twenty-four times what it had been at the beginning. Immediately after the death of Seti I, the clouds gathered; the work in the thin veins of gold-bearing ore became harder and harder – the gold was running out. People began to covet the gold lying in the tombs. Soon unscrupulous individuals, shortly followed by the state itself, began to violate and plunder the tombs.

---

44. J. Yoyotte, 'Stèle de Ramsès II à Keswé et signification historique', B.S.F.E., N° 144, March 1999, p. 58.

The perils from beyond the frontiers increased considerably: it was no longer just a matter of driving back a few 'miserable barbarians' to the frontiers of the only state in the world at the time. Peoples from Assyria, from Babylon, from the Hittite kingdom or Mittani and, later, from Nubia and Libya, would cast their eyes on Egypt. They all came from arid regions and coveted the fertile plain: the name of Egypt is cited six hundred and twenty times in the Bible. It is seen as the country of stable government but also as an impious land of milk and honey. The attackers returned to the assault again and again.

In the words of the French poet Charles Péguy, 'Everything begins as mysticism and ends as politics'. The situation in Egypt was becoming difficult to control, with provincial governments encroaching more and more on the Pharaoh's prerogatives, and the Pharaoh's foreign policy was complicated by uncontrolled immigration through excessively permeable frontiers.

At Abydos, Seti I, full of confidence in the future of his country, had erected the pillar of Djer as a symbol of stability. The pillar is now cracked and the empire itself was breaking up: the south was too remote from the north, and the discipline necessary to overcome the handicaps of geography was lacking.

The New Kingdom was weakened by a crisis in values. In his book *Affaires et scandales sous les Ramsès*, Pascal Vernus shows that strikes, embezzlement, corruption at the highest level and indecent behaviour were already an everyday occurrence.

Owing to its very special geography, the country depended for its equilibrium on the iron fist of Pharaohs determined to maintain a just balance between north and south. This is borne out by the history of its successive capitals. Memphis was too far north yet claimed to be the 'Pivot of the Two Lands' while Thebes, too far south, was chosen once all danger from abroad had been averted. It was a marvellous hub for communications but commercial ambitions gradually sapped the warrior mentality. When the danger from the Hittites became menacing, Ramesses II constructed Per-Ramesse in the delta as an outpost with no other defences than a surrounding wall in brick and a row of sphinxes to guard the gates of the city. When the branch of the Nile on which it was located dried up, it was abandoned in favour of Tanis.

The founding of these new capitals simply added to the resentment of Thebes, whose high priests became jealous of the religious rites performed elsewhere. There was no longer a state like at the beginning of the Old Kingdom, powerfully . centred on the person of the King; the priesthood fomented intrigues and division.

The warrior spirit gradually declined, yielding to a certain spinelessness and shortsightedness. The passionate Thutmose III, the Napoleon of ancient Egypt, said: 'I have cut to pieces its cities and its tribes. I have set the land on fire. The land has been left desolate, transformed into smoking ruins where no tree grows'. He was succeeded by ostentatious or mystical Pharaohs, more concerned with the building of royal palaces than with the defence of the country. The army was demoralized by inaction or by being made to work in the quarries or on the Pharaoh's construction sites. It neglected its martial values and recruited increasing numbers of undisciplined mercenaries. Then came some more realistic Pharaohs – the general Horemheb and the first two Ramesses – but even Ramesses II preferred to counter the Hittite threats by skilful diplomacy rather than by punitive expeditions, while at the same time celebrating his half-victory in dithyrambic terms. During a reign that went on too long, he ceded to the ruinous pleasure of building the most sumptuous monuments while his empire deteriorated. At his death, 'pirates' were landing on the coast of the delta and a Libyan king was invading from the west.

After Ramesses III, who was assassinated, the country fell apart. Discontent rumbled among the peasants, the class that produced the country's wealth but was constantly the victim of requisitions, forced labour and all manner of taxes. It seemed as if the goddess Ma'at was becoming increasingly blind and the Theban priesthood was working behind the scenes to stir up trouble; later it would defy the royal authority and try to take the place of the king. At the same time social movements sprang up and, in 1156 BC, because wages had not been paid, West Thebes was shaken by strikes.

At the beginning of the first millennium BC, Shoshenk I, a Libyan, became Pharaoh. Two hundred yeas later it was the turn of a Nubian, Piankhi. Shortly afterwards the Assyrians invaded the country and sacked Thebes, driving into slavery the entire population and the 'Divine Adoratrice', last of the successors to the high priests.

The Saïs kings, so dear to the heart of the goddess Neith, took the situation in hand before the first Persian invasion. Like the great Egyptian Pharaohs, they restored order and gave the country a period of prosperity during the seventh century BC.

With Nectanebo II (Thirtieth Dynasty) ended in 340 BC the long list of Egyptian Pharaohs. Sensing this, he erected obelisks only a few metres in height, much more modest than those of his predecessors and carved out of a more humble schist.

GRADUAL FADING OF THE MEMORY

The Ptolemies settled in Alexandria in accordance with the wishes of the conqueror: 'Alexandria ad Egyptum' – 'Alexandria near Egypt' – as Strabo called the city: it became a Greek city within the confines of Egypt. The Ptolemies would, so to speak, turn their backs on the country, requiring it only to finance their expenditure. The Nile became a river without a soul, a mere hydraulic system whose only purpose was to produce wealth.

But the first act of the early Ptolemies, who, like Alexander and before him the Persians, recognized the power of the priesthood, was to visit the major shrines. Indeed, in 276 BC, Ptolemy II contravened Greek customs in his desire to conform to the image of the Egyptian Pharaohs and married his own sister, who became the goddess Philadelpha 'she who loves her brother'. She was soon worshipped in all the temples of Egypt.

Policies dictated by circumstances need to be distinguished from history. At the time of the Ptolemies, the memory of Egypt was founded on the respect for the past cultivated by the priests. The precision of this memory greatly impressed the Greeks and in particular Plato who, in his *Timaeus*, makes an Egyptian priest say the words we have already cited concerning their respect for the past. Plato's admiration was perhaps excessive and his words are surprising coming from a philosopher whose nation had made a goddess of memory: in ancient Greece, bards capable of memorizing thousands of lines of poetry were highly esteemed.

Ptolemy Philadelphus, intelligent and cultivated, showed the importance he attached to memory by founding the famous Library of Alexandria, which sought to assemble the knowledge of all countries under one roof. At this time Alexandria was a city of two languages, two cultures and three systems of writing. It was this King who asked Manetho, an Egyptian priest, to set down in written texts the entire history of Egypt.

Before this monument of history disappeared, it inspired many of the leading Greek scholars living in Alexandria. Ptolemy also tried to produce a map of Egypt and Eratosthenes journeyed as far as Aswan to verify that, at mid-day on the summer solstice, the rays of the sun penetrated right to the bottom of a well and on his return to Alexandria he deduced from this the radius of the earth. Euclid later wrote a treatise on proportions but in this case he plagiarized the work of the Pharaohs, forgetting that the harmony of the pyramids had already expressed these astonishing proportions.

It was the great age of intelligence and of the preservation of acquired knowledge. But Alexandria would rapidly become a world of ostentation and pleasure, with each succeeding Ptolemy yielding to the desire for an ever greater show of wealth, though they did build a few temples – at Edfu, Philae and Dendereh – in the purest Egyptian style. Strabo suggested the words of Homer – 'Each engendered the next' – as an appropriate description of the Ptolemies. This reference to Homer was unfortunate as, after the half-century on the throne of the first two Ptolomies, their successors sank into infanticide, parricide and murder, married their sisters or their daughters, and so on. When Strabo says they impoverished themselves to build their palaces, we should not feel too sorry for them: they had built up an army of zealous functionaries skilled at squeezing the peasants and bringing back gold to Alexandria. The distant successors of Alexander had lost the true memory of the Pharaohs, which they sometimes travestied. One Queen, for example, awarded herself the absurd title 'Female Horus, sovereign of the Two Lands, mighty Bull'.[45] All these later Ptolemies with their various nicknames

---

45. Suzanne Frain, *L'Inimitable*, Fayard 1998.

– 'bighead', 'chickpea', 'flute-player' – remind us that the world is often ruled by people quite different from what their subjects imagine.

The Nectanebo II of the Greek Pharaohs was Cleopatra VII. Horace sums up her character in two harsh words, 'fatale monstrum', while Shakespeare called her a 'triple strumpet' and 'the devil's mare'. This is a very narrow view since the Queen, intelligent and well-educated, had, unlike some of her ancestors, a highly developed political sense; she had certainly read the book of Manetho on the history of Hatshepsut and she was just as ambitious. I suspect that she had even read the story of Rhodope, the great courtesan, who was, Herodotus tells us, 'so famous that every Greek was familiar with her name'. Her charm was less gentle than that of Rhodope – she rewarded Caesar by murdering her own husband Ptolemy XIII and becoming his mistress before becoming the mistress of Mark Antony.

Queen with a tragic destiny, her death marked a turning point in the history and memory of Egypt. Before engaging battle with Mark Antony at Actium, Octavian, a subtle politician, had aroused in the Romans a hatred of Alexandria, city of vice, and of the bewitching Cleopatra. After her death, the country ceased to defend itself against the Romans. 'To guard the country', wrote Strabo, 'the Romans needed only three cohorts, and not even at full strength'. The new conquerors were mainly interested in the collection of taxes and the production of wheat. Paradoxically, though they were little inclined to be religious, they became fascinated by the Egyptian rites of mummification and by their belief in immortality – and this would eventually give rise to the mysterious portraits found in Faiyum.

The memory of the Pharaohs was fading. Vitruvius, a devoted follower of Octavian, now the Emperor Augustus, makes not the slightest mention of the Egyptian temples in his *De Re Architectura* and even, in his seventh chapter on the 'mirabilia aquarum', makes the blunder of situating the source of the Nile in Mauritania! Pliny, who had not travelled, felt himself duty-bound to criticise the pyramids, '*a pointless and idiotic display of royal wealth*', and then, placing Herodotus on the same level as a dozen obscure writers, observes that they do not all agree about the names of the builders of these monuments and concludes: '*It is the most just of destinies that the authors of such a vanity be forgotten*'. He forgets that, more than four centuries earlier, Herodotus had, unlike the others, correctly identified the builder of the Great Pyramid. We have already seen the bias of Horace. Historians, then as now, were not tender towards their fellow historians: Strabo makes a point of emphasizing that Herodotus liked telling tall stories, but his own *Geographia* is full of mistakes.

From now on the past would fall more and more into oblivion. The last hieroglyphic inscription dates from 24 August 394 AD, during the reign of the Emperor Theodosius. The followers of the new religion were no longer tolerated and Neith was right to complain about the destruction ordered by the Emperor Theodosius. A great many temples were laid waste or their ruins turned into houses or churches. The Pharaoh of the ancient times, interlocutor of the gods, was no more.

Sultans would send for lime-burners to recover the limestone blocks that had been the glory of the Egyptian monuments. The Lord of Anglure, a French aristocrat who made the pilgrimage to the Holy Land in 1395, visited Giza and saw for himself the vandalism taking place before his very eyes: from the top of the pyramids 'they let rush down to the bottom their most noble masonry'. It was deliberate destruction since 'the Sultan takes two parts of the profit from these blocks that are thrown down and the masons one third'.

Desert sandstorms did the rest, but certain monuments were saved by the very violence of those winds, which wrapped them in a winding sheet of sand. Such was the case with the Temple of Abu Simbel and the Sphinx. After repairs made even in ancient times by the Greeks and Romans, the latter was envelopped in a greatcoat of sand.

To this destruction by men and by time, the Nile added its own twists: when Jollois and Devilliers, two savants brought to Egypt by Bonaparte, went into the pillared hall of Karnak, they observed that some of the splendid columns were no longer vertical owing to the perverse effects of the Noun, the infiltrating waters of the Nile, which had finally had its way with them.

Part 2

CHAPTER 3

The nineteenth century:
Persistence of the Pharaonic dream

Being at a strategic crossroads, Egypt has always fallen prey to empire-builders attracted by its reputation for abundance. It was ruled by the Romans, then by the Orthodox Christians of Byzantium, and later by Arabs, Mamelukes and Ottomans. It was not until the nineteenth century that the country had its first dreams of independence.

EGYPTOMANIA AND PHARAONIC AMBITIONS

A handful of travellers revealed for Europeans the strangeness and the beauty of the country's monuments. In the Age of Enlightenment, what a paradox it was to see, amid the bubbling of ideas, the pure lines of the pyramid, an architecture that had once embodied the royal power of divine right. The fascination for Egypt began. The pyramid became the symbol of a new order; it appealed to the members of secret societies because its four sides converging to a point seemed to enclose a secret. Masonic initiation rites were held in places shaped like pyramids and the society of reformers dreamed of mysteries. Queen Marie-Antoinette herself went with the Court to the Desert of Retz to admire a stone pyramid that had recently been built.

After the overthrow of the monarchy, people would go to the festivities on the Champ de Mars – where the Eiffel Tower now stands – to see the pyramid erected in honour of the defenders of the homeland, with its inscription 'Tremblez tyrans! Nous nous levons pour les venger.' Egyptomania took hold of the popular imagination and the Pharaohs were depicted as great liberals. Collectors of all kinds milled about and genuine artefacts or forgeries, objects of historical importance or mere trinkets were offered for sale.

Much less inoffensive was the dangerous illusion of pharaonic ambition. An ageing monarch or a future conqueror would see himself as a mighty Pharaoh who, with absolute authority and the power of reason, wished to surpass ordinary mortals and be remembered for ever. His dreams would focus on the Great Pyramid and the impressive temples, those 'monuments of eternity'. This form of megalomania appeared in the nineteenth and twentieth centuries on the plain of Egypt – where it was called *fir'aùnia* – and elsewhere.

BONAPARTE

The post-revolutionary ideologists, the people who shaped public opinion, were in favour of a scientific expedition to free Egypt from the yoke of the Mamelukes. To them, a society was despotic when not governed by reason. France seemed oblivious of the fact that the Pharaohs were often both despotic and rational. The arguments advanced by the French were in fact a pretext that masked a conquest in the colonial tradition: Talleyrand, Minister for Foreign Affairs, stated in a Memoir to the Institute that Egypt would be 'a colony as valuable on its own as all those that France had lost'.

The expedition was placed under the command of General Bonaparte, who was very popular at the time, indeed too popular in the eyes of the Directoire. In his view, Egypt and the Pharaohs were one: in his *Memoirs*, he expressed his admiration for their organizing genius and even, in some cases, for their military genius. No doubt he, too, wanted glory. He knew that Egypt had brought glory to the Pharaohs and to certain conquerors of the country. 'We are twenty-nine years old, the same age as Alexander', said Bonaparte to his comrade Bourrienne. Perhaps he too was fascinated by Thutmose III, now sometimes called the *Napoleon of the Pharaohs*. He was certainly attracted by the East, like a number of great conquerors, but just as important for him was a political calculation: to keep out of the way while the Directoire made itself unpopular.

Alexander had been welcomed as a liberator by the Egyptians; not Bonaparte. The Egyptians have never ceased to regard him as an invader.

At the dawn of the twentieth century, Egypt was still dominated by the Ottomans, but the power of the Pasha, the representative of the Sultan, was undermined by quarrelling beys and Mamelukes; anarchy held sway in the country while abroad the Sublime Porte had to contend with Russia, Austria and two other colonial powers, England and France.

The foolish and extravagant expedition of Bonaparte is too often remembered only for its cultural aspect. Even today, the horror of the destruction of the French fleet at the Battle of Aboukir Bay, the number of sailors sacrificed,[1] the sacking of the El Azhar Mosque, the bombardment of Cairo, the massacre of two thousand five hundred prisoners at Jaffa,[2] the administration of opium to pest-ridden soldiers with whom Bonaparte did not wish to encumber his retreat and who would be massacred by the Turks, the cruelty of Bonaparte who, to inspire terror, said 'Every night we shall have thirty heads cut off', the future First Consul abandoning his army without apparent reason (Kleber would be obliged to explain to the

---

1. Eight hundred dead, including the Admirals Brueys d'Aigaïllers and Dupetit-Thouard who, horribly wounded, stood on a barrel of bran and continued to give orders to his crew.
2. An eye-witness, the cavalry officer Jacques Miot described the scene: 'A frightful pyramid of dead and dying bodies dripping with blood because, to economise on ammunition, the prisoners were executed with bayonets'.

soldiers that if their general had abandoned them it was for their own good) and, lastly, the misconduct of the soldiers who, according to an Egyptian song, roamed drunk about the city looking for women.

What a contrast between the toast of Gaspard Monge at a dinner offered to leading Egyptian personalities – 'To the uplifting of the human spirit, to the progress of enlightenment!' – and the exactions of the French army, against which some of the scholars accompanying Bonaparte protested. At St Jean d'Acre in Palestine, Louis-Joseph Favier denounced in public the cruelty of the unscrupulous conqueror. At a meeting of the Institute in the presence of Bonaparte, René-Nicolas Desgenettes, chief doctor of the expedition, called the administration of opium to the pest-ridden soldiers a criminal act, and Vivant Denon stated that 'the difficulty of distinguishing enemies by their physical aspect or colour led to innocent peasants being killed day after day' and he preferred to delay his arrival in the villages 'in order not to hear the cries of the inhabitants being robbed'.

There was of course a brighter side to this sombre record, about which French public opinion was largely unaware throughout the Consulate and Empire. This was the work of the scholars at the Institute of Egypt founded by Bonaparte, but its impact was not immediate: the first edition (incomplete) of the *Description de l'Egypte* did not come out in France until 1809 and the full edition until 1823, the same year as Jean-François Champollion's *Lettre à M. Dacier,* which spoke of his deciphering of the ancient Egyptian hieroglyphics. It is unlikely that the *Description de l'Egypte*, which was particularly awkward to handle and difficult to consult,[3] was a best-seller among the Egyptian intelligentsia who, after the sacking of the El Azhar Mosque cried: 'On this night, the army of God the Merciful gave free rein to the army of Satan'.[4]

If we take a close look at the minutes of communications to the Institute of Egypt,[5] we find that Monge, Fourier and Geoffroy Saint-Hilaire presented a number of learned papers that had nothing to do with Egypt[6] and that the naturalists plunged enthusiastically into the study of fish.[7] At the time, there was in Egypt a sharp rejection of French culture: the Egyptians did not even raise their eyes when the Montgolfières – hot air balloons – of Conté rose into the sky. 'The people',

---

3. Many of those who speak of it have not had it in their hands; it has to be kept in a special cabinet and it requires a lot of patience to consult for there is neither a table of contents nor an index. It contains many repetitions and errors, especially in the maps and in the copying of hieroglyphics, but without descending to the sarcasm of Champollion, who called it 'bilge-water', we have to admit that the whole project was an important step forward in our knowledge of Egypt.
4. Allusion to the speech made by Bonaparte on his arrival in Alexandria, which began: 'In the name of God, the Beneficent, the Merciful...'
5. J.E. Goby, *Premier Institut de France*, 1987.
6. Examples include Monge's paper on a curved surface in which all the perpendiculars are tangents to the same sphere and Geoffroy Saint-Hilaire's paper on the history of the formation of eggs.
7. On reading a paper by Geoffroy Saint-Hilaire, the secretary of the diwan Al-Mahdi expressed his astonishment that someone could write so much about a single species of fish when the Almighty had created fifty thousand.

cried Dolomieu, 'have neither curiosity nor the desire to emulate: their utter indifference towards all that is foreign to their state, profession or customs is perhaps, to my mind, the most extraordinary aspect of their way of life'.

Poor Monge! President of the Institute of Egypt, he enjoyed the full confidence of Bonaparte, whom he followed faithfully to Jaffa and St Jean d'Acre. It was during the exhausting retreat in the desert that he discovered the explanation of mirages! The scientific spirit never slumbers, but by what mirages has it not been taken in? What disillusions he had suffered since leaving the cherished school he had founded on the slopes of the Mont Sainte Geneviève and whose motto was 'For France, for science and for glory'. Alas, the horrors of that war and the death of his close friend Caffarelli were no mirages. The latter had lost a leg but had nevertheless become a general in command of the eastern army's engineering corps. Monge brought back the blood-stained sketch of the defences of the impregnable fortress of St Jean d'Acre made by the general. After fourteen attacks and two months of effort, Bonaparte had been obliged to raise the siege and, having lost a third of his army, he returned to Egypt via Pelusium, which was once the main point of entry into Egypt for menacing hordes from Asia. Would Bonaparte remember this later when in 1808 he awarded Monge the strange title of Comte de Péluse, or should we see it as a reminder of the northern end of the canal that both of them had cogitated? On his return to Cairo, Bonaparte hosted a dinner and announced that the fortress of Acre had been razed to the ground, with no stone unturned. Later, he encouraged the development of numerous emblems to his own glory as a conqueror (Fig. 60).

Figure 60. The Battle of Cairo took place at Imbaba and not at the foot of the pyramids. It was fought at the height of summer in torrid weather – one hopes that the grumblers in the army had unbuttoned their woollen uniforms. It was the start of the Napoleonic legend. (Photo Bibliothèque Nationale).

For the other side of the picture, it is worth lending an ear to what the Egyptians of the time had to say, in particular Abd al-Rahman al-Gabarti (1754-1824), whose book *Histoire du temps passé par les Français en Egypte* contains an account of two revolts, in October 1798 and in March 1800, and of the harsh repression by the occupying forces, with details of their crushing requisitions, the obsession of the soldiers to get themselves women and the various scandals provoked by the misconduct of the French.

At Saint Helena, Napoleon harked back again and again to his memories of Egypt. He was only too well aware that not everything had been perfect on that expedition. In 1810 he had ordered the destruction of a number of documents. At Cherbourg in Brittany, whose roadstead he wished to protect with an impregnable wall, he had the words 'J'avais résolu de renouveler les merveilles de l'Egypte' (I had resolved to recreate the marvels of Egypt) engraved on his statue and, like a true Pharaoh, he dictated at Saint Helena an autobiography[8] openly economical with the truth: 'The departure of the commander-in-chief... was part of a truly magnanimous plan... One can only laugh at the stupidity of those who consider his departure as an evasion or desertion'.

A COMMEMORATION A QUARTER OF A CENTURY AHEAD OF TIME

In these circumstances it is hardly surprising that Egypt has formally refused to be associated with the celebrations of the bicentenary of Bonaparte's campaign. 'I have never heard of a people commemorating the arrival of an invader on its territory', wrote the Egyptian journalist Mohamad Heykal recently. It was also pointed out in Egypt that the shortness of the campaign ruled out any real cultural exchange with the Egyptian people.

In contrast, the French were reminded that when the Arabs withdrew from Spain in the fifteenth century, they left behind unmistakable evidence of their great culture, and yet that gave no cause to celebrate.

The expression 'Bicentenary of the expedition of Napoleon Bonaparte' was abandoned at the suggestion of Egypt in favour of 'Shared Horizons', a form of words suggested by a subtle ambassador which masked big differences in the degree of enthusiasm with which the anniversary was celebrated on the two sides of the Mediterranean. But even in this attenuated form, the commemoration is some decades ahead of its time. If there were a true bicentenary to celebrate in Egypt and in France it would be in a quarter of a century, to mark the revival of the memory of the Egyptian past epitomised by the moment when Champollion unlocked the written memory of the Pharaohs. The scholars travelling with Bona-

---

8. E. Las Cases, *Mémorial de Sainte-Hélène*, Garnier 1961, Vol. I, p. 141.

parte had brought in their harvest and had revealed to the West the immensity of the pharaonic heritage, but that heritage had remained unintelligible.

As André Raymond so rightly states:[9] *'A historical narcissism, pushed to the extreme by the exaltation of the 'Napoleonic epic' and the overvaluing of the scientific results of the expedition led the French to exaggerate the positive impact of the expedition and its results and to go so far as to date the awakening of Egypt at 1798'.*

AWAKENING OF NATIONALISM

There had been invasion by the French and resistance by the Egyptians. Even though the campaign did not immediately awaken Egypt to French culture, its direct consequence was that the inhabitants of the ancient land of the Pharaohs discovered the idea of nationhood. And this idea developed, with ups and downs, throughout the nineteenth century. As early as 1788, Constantin de Volney (1757-1820) had predicted that 'to gain a foothold in Egypt' three wars would have to be fought, the first against the British, the second against the Sublime Porte and the third, the most difficult, against ' the Muslims who form the population of that country'; and it was there that Bonaparte failed because of the profound difference in customs and religion. The Egyptians never fell for Bonaparte's alleged sympathy for the Koran: indeed, they sensed the danger of the obliteration of their religion by the 'Christian' Bonaparte. Danger from abroad and the insulting of their religion formed a most effective cement for creating a sentiment of patriotism. Bonaparte had failed to learn some lessons from history: Cambyses and Darius, the Persian conquerors, had proved respectful masters of Egypt. 'When King Cambyses came to Saïs he went into the sanctuary of Neith. He worshipped before the holiness of Neith with much devotion, as all the kings had done, he made great offering of all good things to Neith, the great, the divine mother, and to all the gods who dwell in Saïs, as all the pious kings had done'.[10] Later, Alexander had been the protector of the Egyptian religion and had gone to the oasis of Siwa to consult the oracle of Ammon.

MEHEMET ALI, A SELF-TAUGHT FOREIGNER
WITH TRAITS OF A PHARAOH

When it pulled out of Egypt, the French army left behind nothing but hatred. Its leader, general Menou, had to surrender to a Turkish-British coalition. The Turkish forces of Selim III were placed under the command of an Albanian called Mehemet Ali. The latter watched in silence as dissension broke out between the

---

9. A. Raymond, *Egyptiens et Français au Caire, 1798-1801*, Cairo IFAO, 1998, p. 366.
10. W. M. F. Petrie, A History of Egypt, London, Methuen 1882.

Turks and the British, who evacuated Egypt in 1803, leaving the country in the deepest disorder. He then proceeded methodically to eliminate his opponents, and he did it so effectively that two years later, in 1805, the Sublime Porte ratified his appointment as Pasha of Egypt.

This man was to rule Egypt for forty-four years. An exceptional figure, he left a profound mark on the country, worthy of a Pharaoh. Oddly enough, however, he never tried to imitate any of those ancient kings: not knowing how to read or write, he took little interest in the past. He was a man of action.

In the first place he freed the country from the heavy-handed Ottoman yoke. After his brilliant victory over a coalition of Turks and British in 1807, the Sublime Porte came to rely on the new master of Egypt.

Mehemet Ali, it seems, had his own dream of the Orient and, when he had freed the Holy Land, he found himself on the banks of the Euphrates, just like the great Pharaoh Thutmose I. Like that Pharaoh, he craved to dominate the country. In 1808 he declared himself owner of the land and its people. He showed little interest in the lives of the fellahin – he wanted to make the country a producer of riches. Like him, the Pharaohs had not been kind to the 'human cattle': they had directed the energies of the people to the construction of immense and beautiful royal projects, but Mehemet Ali used them to increase his personal wealth, provoking R. Madden[11] to say in 1841 that 'a more rapacious ruler never reigned in Egypt'. The judgment was, like those of several of the Pasha's contemporaries, much too severe, for he used most of the money to improve agriculture, found an army and navy, dig canals and build roads, factories and schools, etc.

He was diplomatic, subtle and cunning, in the words of Robert Solé,[12] 'a true political animal'. He spent his whole life dividing to rule – driving wedges between the various segments of Egyptian society or between the great powers, the Porte, England and France. When he feared that the struggle might last too long, he negotiated: he was 'capable at any moment of obtaining the support of France to refuse a British demand and of obtaining that of Britain to oppose a French scheme'. His political instincts were reminiscent of Ramesses II in his negotiations with the Hittites.

And then, to modernize Egypt and improve its productive capacity, he did what Ahmose,[13] the Saïs kings and the early Ptolemies had done: he opened up the country to foreigners, to Greeks, Armenians, Syrians, Italians and French. Within this foreign influx the French gradually came to play an important part, for several reasons. Firstly, on the advice of Drovetti, the French Consul in Egypt, France had played the Mehemet Ali card against Britain. The glorious feats of the Napoleonic saga were used to win him over and efface the disastrous impression left by Bona-

11. R.R. Madden, *Egypt and Mohammed Aly.*
12. R. Solé, *L'Egypte, passion française*, Seuil 1997.
13. At Avaris have been found frescoes showing the recruitment in the time of Ahmose of Minoan craftsmen.

parte. Mehemet Ali forgot the candidate First Consul and dreamed of imitating the Emperor. In France, the Restoration had not been to the taste of all the French: Monge, ever faithful to the Emperor, had been stripped bare and even hounded from the Institute; in France, too, there were quite a number of old soldiers from the imperial armies in disgrace who hoped, like others, to find employment abroad. Thus it was that Coste arrived in Egypt in 1817, Sève in 1819, and Clot in 1825. Coste would have an important influence in architecture; Sève, an officer of the Grande Armée, became Soliman Pasha, organiser of the Viceroy's army, and Clot became Clot Bey, personal physician of the Viceroy. They joined the young Linant de Bellefonds[14] (1800-1883) who had entered the service of the Pasha at the age of eighteen and rose to become his chief engineer before ending as Linant Bey, Director of Bridges and Roads under Saïd Pasha. It was he who would rediscover the memory, lost by the Mamelukes, of the subtle water management system that had made the plain so rich – the dykes he had built recall those of the Pharaohs.

But it must not be believed for a single instant that the French were privileged: the key ministries were always held by orientals. The choice was made after a rigorous selection process: '*I know that, out of fifty individuals who come to offer me their services, forty-nine are like fake precious stones. Without testing them, however, I am unable to discover the single true diamond amongst them. I begin by buying all of them and when I discover the true stone it repays me a hundredfold for the losses I incurred with all the others*'.[15]

In his desire to create a modern state, the Viceroy sent Egyptian students to Italy as early as 1809 but in 1826, on the insistance of Drovetti, four young men made their way to France accompanied by the Imam Rifaa el-Tahtawi. The lattter, a great observer of French civilization, published a book entitled *L'Or de Paris, Relation de voyage* (The Gold of Paris, Account of a Journey) in 1834. Mehemet Ali ordered it to be circulated widely and established numerous schools in which French teachers played a leading role. It was now, twelve years after Champollion's '*Letter to Monsieur Dacier*', that one could truly speak of 'shared horizons', but the West-East current would always predominate.

Mehemet Ali, in sharp contrast to the ancient inaccessible Pharaohs, wanted contact with foreigners. He was liberal in spirit: before he came to power, obscurantism was rife in Egypt. Even at the El Azhar Mosque almost nothing outside religious subjects was taught. The teachers would assert that the Pharaohs worshipped idols and stress their persecution of Moses. Mehemet Ali was completely indifferent to this, for two reasons.

---

14. He is the most important source for the history of construction works in Egypt; his book, *Histoire et description des principaux travaux d'utilité publique exécutés en Egypte depuis la plus haute antiquité jusqu'à nos jours* (History and description of the major public works constructed in Egypt since the most ancient times to the present day), published in 1874, is not easy to read but is of great interest.

15. A.G. Politis, *L'Hellénisme et l'Egypte moderne*, 1928.

In the first place, he lived intensely in the present and was totally uneducated: he did not learn to read and write until he was forty-seven years old. Ancient architecture was of little interest to him and he regarded all that stonework as quarries for the dams he was building in the delta: Linant de Bellefonds tells us – and his account gives a glimpse into the ruler's mind[16] – how he managed to dissuade the Pasha from demolishing the whole of the pyramid of Mycerinus and part of the pyramid of Cheops to build his dams. Champollion played no part in this decision. The Viceroy only gave up his project when his minister came up with a sordid calculation of the relative costs: delivery of the stone from the pyramids to the dam would cost 10.20 piastres as against 8.25 for stone extracted from the usual quarries! Champollion emphasized the archaeological value of the obelisks but the Viceroy, as materialist and calculating as ever, saw them simply as a means of payment for services rendered by another country. Owing to his lack of culture he did not realise that the hieroglyphs engraved on those obelisks showed how much importance his distant predecessors attached to written communication.

Secondly, Mehemet Ali had no religious convictions. He took away the land of Muslim establishments in the valley and only agreed, after a long period of reflection, to liberate the Holy Land because of the prestige it would give him.

One other trait of his character should be mentioned – his cruelty. After his victory of 1807, he put the heads of the defeated English on display in Cairo and, after the bloody massacre of the Mamelukes in 1811, he sent their heads to Constantinople. But were not some of the founding Pharaohs of Egypt just as barbaric in certain circumstances? One example is given in Figure 61.

His son Ismail would pay with his life for the cruelty and greed he had inherited from his father. He was coming to the end of a pitiless hunt for slaves among the populations in the east and southeast of present-day Sudan when, passing through Nubia, he stopped at Chandi, a town of fifteen thousand inhabitants. There he demanded two thousand slaves and a tribute equivalent to some twenty thousand gold francs. The local chieftain Nimr refused. Ismail struck him and threatened to have him impaled. Nimr appeared to give way and even invited Ismail to a sumptuous dinner. But Nimr had ordered the hut in which the dinner was to take place to be surrounded with hay. At the end of the dinner the hut was set on fire and Ismail was burned alive. To avenge him, Mehemet Ali sent one of his sons-in-law known for even greater cruelty.

Heavy taxes were imposed by the Viceroy on the fellahin. The names given to these taxes reflect the ingenuity of the ministers for finance: there was first the 'good news tax' and then the 'solidarity tax'. Under the latter, the neighbour of a fellah unable to pay his taxes had to pay them in his place and other neighbours should the first one fail to cough up, and so on frcm village to village. The Viceroy was also so wily in business that he bought up cheaply all the beans (staple

---

16. Linant de Bellefonds, *op. cit.*, pp. 420-424.

Figure 61. The Pharaohs were not tender-hearted towards their enemies. This palette shows standard-bearers and beheaded prisoners preceding the king. Myth or reality? It is an open question, but with Mehemet Ali there can be no doubt.

food of the people) in the country and later resold them himself, forcing the fellahin to buy back their own production at a high price.

The tax policy of Mehemet Ali had a purpose: major public works undertaken on his orders would arouse Egypt from its lethargy. At the time of his accession to power, 'each village, administered by a caimacan, organized the digging and maintenance of the canals and dykes for the irrigation of its own land, without worrying about whether they were acting for or against the interests of their neighbours. There were constant squabbles, often followed by skirmishes, between the various districts. It was under Mehemet Ali, in around 1816, that work was started on digging major canals; the river was hemmed in between powerful embankments and dykes along its length... a mighty project... in which up to 50 million m^3 of earth were shifted in a single year'.[17] The Viceroy made up for the negligence of the Mamelukes and returned to the policies of the foremost Pharaohs.

In an industrial century which saw the first appearance of the word 'socialism', Mehemet Ali embodied a state socialism in the hands of one man. Linant de Bellefonds, who was devoted to his master, put it like this: 'Egypt cannot be compared to any other land. It can be likened to a large farm of which the viceroy, the head of the state, is the head farmer. It is he who has to direct it and make it pros-

17. Linant de Bellefonds, *op. cit.*, p. 118.

per for the benefit of everyone, and everyone in this immense rural undertaking must work to produce as much as possible. It is as it were an immense phalanstery in which each individual is expected to work for the good of all and shares in the results obtained'. Linant de Bellefonds became too wedded to the interests of his master to ask himself whether everyone did in fact really profit from the results obtained. He was later forced to admit that it was not always possible to reward each person for the work they had done.[18] He even dared to suggest that forced labour was accepted willingly, even though he knew perfectly well what it all meant: *'The workers leave their villages and their work on their farms, to dredge, without any payment, a canal that is of no benefit to them; they must bring with them their own tools, and the women and children follow the head of the family as is the custom... When a fellah is on the corvée, his wife, his children, his father, his mother, etc. go with him and everybody sets up their living quarters unprotected from the sun and wind, on the site of the work to be done'.*

In the nineteenth century these peasants were often beaten and mistreated. To facilitate recruitment at the start of the Suez Canal project, the Canal Company put up notices in the villages that no beatings of any kind would be allowed. Did the elderly Socrates really get it wrong when he asserted that, by the will of God, a thankless fate, such as the life of a fellah, lay in store for countless generations?

The work on the canals, especially the periodic dredging, was particularly unhealthy: 'The fellahin waded into the slime up to their chests, carrying the dredged mud and sand in sacks on their backs, down to an average depth of 12 m'.[19] It is thus easy to understand the high death rate: 20,000 died during the excavation of the Mahmoudieh Canal under Mehemet Ali and, if Herodotus is to be believed (Book II, 158), 120,000 in the digging of the canal of King Necos, who reigned from 610 to 595 BC.

Moreover, Mehemet Ali was a difficult master to serve. He was always full of ideas and wanted everything to be done at once. Sometimes he was overambitious and sometimes too meddlesome. He abandoned his project of using the pyramids as a quarry, the digging of artesian wells and the construction of cotton mills powered by the cataracts; some of his canals rapidly silted up and the dams in the delta were not a great success – and the same can be said of some of the factories he started. But despite these failures and his personal faults, the balance is clearly in his favour: Mehemet Ali dragged the country out of its torpor and it became aware that it was a nation. When the pupil admires his teacher, he sometimes ends up by adopting some of the teacher's traits. Mehemet Ali was almost a contrary example: he was oblivious of the history of the Pharaohs but he ended up by resembling some of them, which is one reason why so many aspects of his reign are interesting to study.

---

18. *Ibid.*, p. 42.
19. *Ibid.*

DE LESSEPS, CONQUEROR OF THE ISTHMUS
AND VICTIM OF *FIR'AUNIA*

Mehemet Ali fascinated the Saint-Simonists. The Pasha seemed to draw inspiration from their doctrine of state socialism with mobilization of the people around major projects of use to the community. In the eyes of their chief 'Father' Enfantin, the piercing of the Suez isthmus was one such project: 'It would not be just a technical exploit but would respond to a sacred need. To add to the map of the world that blue line would be a sign of peace, concord and love between the two continents'.

But the Viceroy paid no attention to the project, showing greater interest in the railway from Suez to Cairo proposed by the British. The canal would not be built until the reigns of his second and third successors.

VISIBLE AND INVISIBLE MYSTERIES OF THE ISTHMUS

The isthmus has a long history. It facilitated relations between the chosen people and Egypt in the first two millennia BC; it was the route taken for the invasion of Egypt by the Hittites, the Babylonians, the Assyrians and the Persians; here passed the armies of Ramesses II, Cleopatra, Alexander the Great, several Roman generals, Saladin, Bonaparte and, much later, the soldiers of the Six-Day War and the war of October 1973.

The isthmus covers a substrate whose even more tormented history is unique in the world. It is a point of permanent conflict between the gigantic forces unleashed in the collision between the tectonic plates of Africa and Eurasia, the cause of murderous earthquakes, as if Africa wished to punish those who had deserted it. In 1675 BC, the island of Santorin disappeared in a cloud of ashes that veiled the sun as far as Egypt.

Clearly, the eastern branch of the Nile has dried up: Pelusium, formerly a city of mud,[20] has become a tell. Further west, Per-Ramesse was abandoned in favour of Tanis. In a kind of tilting movement (Fig. 63), the western part of the delta dipped as the eastern part was uplifted: the level of the Canope branch of the Nile has dropped several metres and, in the port of Alexandria, rocks that were visible in the time of Strabo[21] have disappeared under water. Under the site of the port of the Ptolemies to the east, research carried out by Frank Goddio from a boat equipped with state-of-the-art electronic apparatus has allowed us to perceive in the murky water the splendid statues and remains of the palace of Cleopatra on the peninsula of Antirrhodos, which was swallowed up by a landslide probably caused by the tsunami of 365 AD, which is said to have projected boats on to the roofs of houses.

---

20. 'Pelusium is surrounded by marshes that some people call Barahra and by stagnant pools of water', Strabo, 21.
21. Strabo, 6.

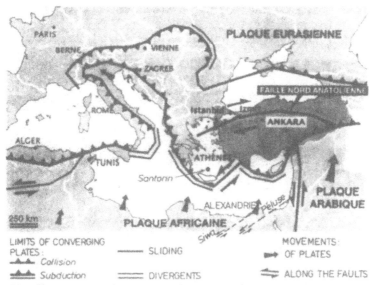

Figure 62. The effects on the surface of the collisions beneath the isthmus are highly complex: specialists distinguish two parallel shear surfaces, Alexandria-Siwa and the fault of Pelusium. It is therefore not surprising that a terrible earthquake occurred on 21 July 365 AD, causing a tidal wave that drowned 5000 persons in Alexandria and long haunted the memory of the Alexandrians.

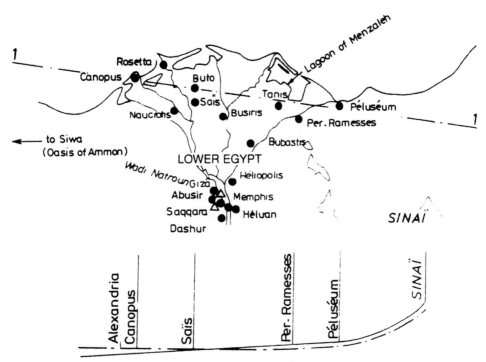

Figure 63. The tilting of the delta along an east-west axis. The broken line shows the land surface several millennia ago, the unbroken one the situation as it is today.

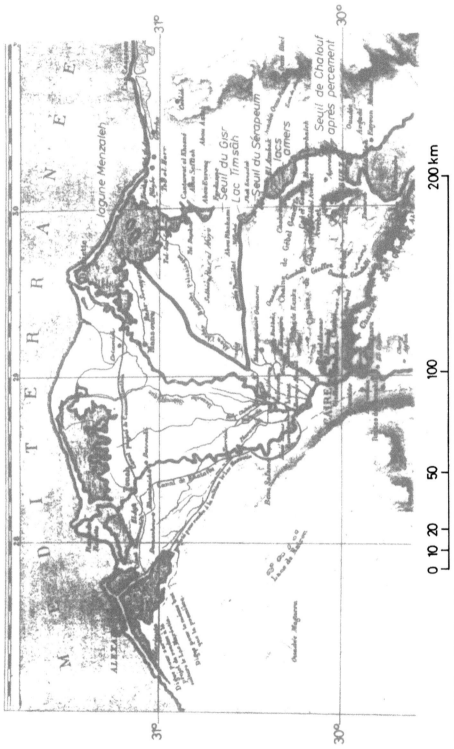

Figure 64. Map of the Suez isthmus in 1858, drawn by Linant de Bellefonds bey. Here and in Figure 68 the word 'seuil' means 'sill', i.e. a low ridge.

The geography of the eastern part of the delta is therefore quite different from what it was at the end of the second millenniun BC, when the exodus of Moses is usually reckoned to have taken place. Today, the isthmus extends from Suez to Port Saïd, a distance of a hundred and twenty kilometres that is punctuated by the two Bitter Lakes and Lake Timsah and, near the Mediterranean, the lagoon of Menzaleh, which has lost three quarters of its surface area; the stretches of water are separated by three sills, the Sill of Chalouf, the Sill of Serapeum and the Sill of Gisr (Fig. 64).

At the time of the Old Kingdom, on the other hand (Fig. 65), the finger of the Red Sea penetrated further northwards: the southernmost sill (Sill of Chalouf) did not exist and the Bitter Lakes were joined to the Red Sea, which extended as far as the Sill of Serapeum, i.e. nearly to Lake Timsah, leaving an isthmus of only 60 km, half of which was occupied by the large Lake Menzaleh. The Bitter Lakes are so called because their waters have remained brackish. This slowly changing geography, a logical development from the very ancient Tethys sea, makes it easier to understand the story of Moses in the Bible and the history of the ancient canals across the isthmus.

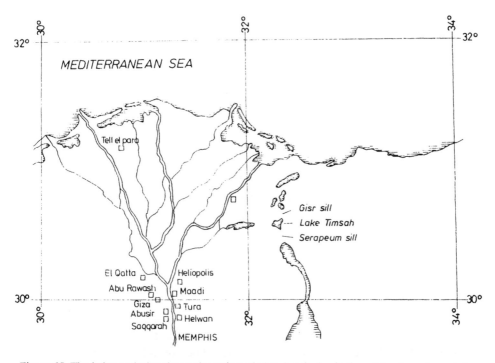

Figure 65. The isthmus during the predynastic period and at the beginning of the dynastic period, according to the Atlas of Ancient Egypt by J. Baines and J. Malek, 1980. The finger of the Red Sea extends well beyond the 30° parallel of latitude.

THE EXODUS OF THE HEBREWS

The two seas bordering the isthmus have small tides but they are very different in amplitude – 0.38 m for the Mediterranean and 1.79 m (with exceptional tides of up to 2.42 m) for the Red Sea. This explains how, in the southern part of the isthmus, in a semi-aquatic area with an unstable soil which was called the 'Sea of Reeds', Moses and the Hebrews were able to escape whereas the pursuing Pharaoh and his soldiers were drowned by the rising tide.

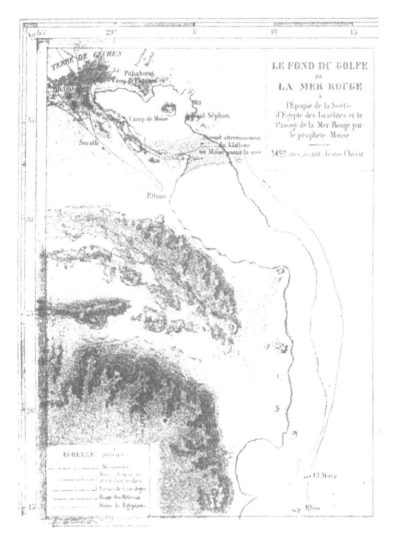

Figure 66. The head of the Gulf of the Red Sea at the time of the exodus from Egypt of the Israelites and the crossing of the Red Sea by the prophet Moses, 1491 years before Jesus Christ. In this map from a little-known study, Linant de Bellefonds, with supreme self-confidence, situates the exodus of Moses in exactly 1491 BC and even marks the site of his camp!

Figure 67. Sinai. In this strange light, it seems to be emerging under the thrust of the African plate (photo from the Fenoyl collection).

Ancient canals from the Mediterranean to the Red Sea

Enfantin, of course, was not the first person to have had the idea of digging a canal to link the two seas. The idea goes back to the time of the Pharaohs and the first actual attempt dates from the reign of the Saïs king Necos. The work was continued and eventually terminated by Darius, Ptolemy II and the Roman Emperor Trajan.

But the canal in question was not at all the south-north furrow from Suez: the plan at the time was to link the Red Sea to Lake Timsah and then to dig a canal in a roughly east-west direction to connect the lake with one of the eastern branches of the Nile. Contrary to what Strabo writes, it was never possible to go 'freely and without impediment' from the canal to the sea. Diodorus says that 'there existed an ingenious device that was opened when they wanted to let a boat through' (Book I, XXXIII, 10). This device was in fact either a dam that could be raised and lowered or a slipway like the ones we have already mentioned, but either of these devices would have been restricted to small flat-bottomed boats. This interpretation is confirmed by some information provided by Plutarch in his life of Mark Antony. Plutarch states that Mark Antony, having arrived at Alexandria shortly before the Battle of Actium, found Cleopatra busy looking for ways of

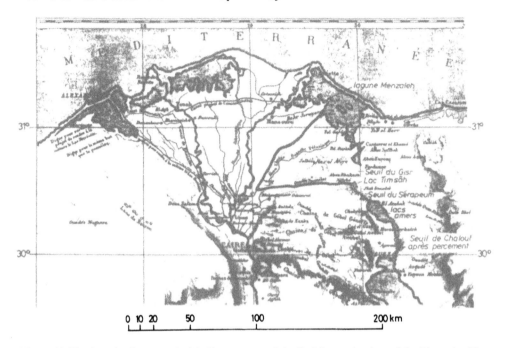

Figure 68. The junction between the Mediterranean and the Red Sea at the time of the Pharaohs. The thick lines, added by the author to the map drawn by Linant de Bellefonds, represent the ancient canals and branches of the Nile that formed the way through.

transporting her vessels across the isthmus separating the two seas so as to be able to escape with all her treasure. In other words, this ancient canal between the two seas could only accommodate keelless boats with a shallow draught; as at Coptos, large vessels were dismantled, transported and then re-assembled on the shores of the Red Sea.

The deeper reason for the 'ingenious device' mentioned by Diodorus was that the Egyptians dreaded that one day the salt waters of the Red Sea might invade the lower reaches of the delta, depriving the local population of drinking water. These fears were not groundless[22] since the eastern part of the isthmus was at the time was a little lower.

CONSTRUCTION OF THE NINETEENTH-CENTURY SUEZ CANAL

In the nineteenth century the problem was no longer finding a safe place for the treasure of Queen Cleopatra but the building of a powerful instrument for international trade.

---

22. Strabo (25) was quite wrong to regard this proccupation of the ancient Egyptians as unfounded.

Barthélemy Enfantin and his disciples arrived in Egypt in 1833, and left in 1835 to escape an epidemic of the plague. Never was 'Father Enfantin' received by the Pasha; the way he had changed the Saint-Simonist movement was already enough to show that he was not the 'genuine precious diamond' the Viceroy was looking for, and it was better not to tell the Pasha about the anti-slavery opinions of the Saint-Simonists, who dreamed of impregnating the negro race, the 'female and sentimental' race, with the 'male and scientific' virtues of the white race in a 'frictionless' society!

For twelve years there would be almost no further talk of the canal. There were two reasons for this. Firstly, the Pasha had awarded a concession to build a railway to an Englishman called Galloway and, secondly, the errors in levelling by the French engineer Lepère during Bonaparte's campaign had left the impression that it would be very difficult to build a canal, since numerous locks would be needed to cope with the difference in level of nearly 10 m (exactly 9.908 m was the figure given by Lepère in his report to Bonaparte) between the two seas. Most of Lepère's findings were not particularly surprising and explained the ancestral fear of flooding mentioned above. Some more thoughtful spirits, however, such as Laplace, Fourier and Monge, found it rather mysterious and those who had heard of the curious gateway between the two seas wondered with astonishment how it could have functioned under such a great pressure of water. The engineer Paul Bourdaloue (of the same family as the famous seventeenth century Jesuit preacher) was therefore called in. He was a self-taught inventor of the system of *mires parlantes* – a new type of levelling pole, so called because it was so much clearer and easier to use – who later became responsible for the surveying of France as a whole. He quickly spotted the unfortunate mistake and further surveys in 1853, 1856 and 1857 confirmed his views.

The canal now appeared less of a pipedream and Enfantin set up in Paris a *Société d'études pour le canal de Suez* which, however, owing to bickering among its members, never came up with valid proposals and, what was worse, impeded the work of de Lesseps.

Ferdinand de Lesseps had been sidelined from the diplomatic corps after an incident and had been seething in the French province of Berry for five years when in 1854 he received news of the assassination of Abbas Pasha, the immediate successor of Mehemet Ali. It was Saïd, the fourth son of Mehemet Ali, who succeeded. De Lesseps had known Saïd at the time he had been consul in Egypt. There was seventeen years between them. Saïd, who had suffered from bulimia as a child, would always remember the kindly welcome he used to receive at the consulate.

De Lesseps arrived in Cairo on 7 November 1854; twenty three days later he was already in possession of a concession granted to him in a personal capacity by the Viceroy. The document begins as follows: 'Our friend, Mr Ferdinand de Lesseps, having drawn our attention to...', and the treaty includes the following main provisions: 'He shall have authority... to constitute and direct a Universal

Company for the piercing of the Suez isthmus among all the capitalists of the entire world... The work shall be executed at the expense of the said Company.... The Egyptian Government shall receive 15% of the net profits independently of the interest and dividends on shares that it reserves to itself the right to buy at the time of issue.... The concession shall be for ninety-nine years and Egypt shall then recover possession of the canal in return for compensation to be fixed.'

This contract, which de Lesseps had drawn up himself and taken to the Pasha for signature, provides a magnificent example of his dynamism, though not of his negotiating skills since Saïd trusted him absolutely. The new Viceroy was far from possessing the realism and perspicacity of Mehemet Ali and he was all too faithful to his friends.

Two Christmases: 1798 and 1854

At Christmastime in 1798 Bonaparte, Monge, Berthollet, Caffarelli and Lepère were all together in the former sea of reeds south of Lake Timsah with the purpose of inspecting the remains of the canal built by the Pharaohs and laying the foundations of a project for a canal across the isthmus. Two years later, Lepère submitted plans to the First Consul for a canal with locks that would reach the Mediterranean at Alexandria: 'It is a great matter', said the First Consul, 'but I am not now in a position to accomplish it'. As things stood, he was quite right to show little interest in the project of Lepère (a former comrade at the military school in Brienne), which was rendered worthless by his enormous error in surveying.

Fifty-six years later, de Lesseps was spending Christmas walking the entire length of the canal's route. He rapidly adopted[23] the short south-north route proposed by Linant Bey and Mougel Bey. As a tribute to the Viceroy, he decided that the point at which the canal entered the Mediterranean would be called Port Saïd. Twelve years later the canal had become a reality in spite of the numerous obstacles that de Lesseps had to vanquish one after the other.

What is particularly interesting in the personality of de Lesseps is the extent to which he epitomized the qualities of a true Pharaoh – authority, decisiveness, determination. He traced in the soil the trench that would separate two continents, a short direct route that no Pharaoh had ever dared to contemplate.

The advantages enjoyed by de Lesseps

To overcome the obstacles in his way, de Lesseps had some very useful connections: besides his great friendship with the present Viceroy and with his successor, he had become, through marriage, the cousin of the Empress of France. In addition, the local geography was very helpful to the scheme:

---

23. In preference to the excessively sophisticated project of Paulin Talabot, a Saint-Simonist, which would have used canal bridges above the branches of the Nile to reach Alexandria.

1. The two seas were at very nearly the same level, as we have said, which made locks unnecessary,
2. The land was almost flat and, with rare exceptions, easy to excavate,
3. The climate was dry,
4. The workforce promised by the Viceroy would be abundant but, if there was a shortage of labour, it might be possible, since the invention of steam engines by JamesWatt, to use some form of machinery.

Lastly, he was able to count on the collaboration of numerous talented French engineers, most of whom had been trained at the *Ecole des Ponts et Chaussées*, the oldest school for engineers in the world, founded by Louis XV. The art of constructing major waterways had long been part of its curriculum and the school still honoured the memory of Pierre Paul de Riquet (1604-1680), the superintendent of Louis XIV and builder of the Two Seas Canal[24] across south-west France from the Mediterranean to the Atlantic, an exceptional accomplishment of magnificent design that boldly incorporated a whole battery of locks to mount the Gate of Narouze, the 192 m watershed between the Atlantic and the Mediterranean. At that spot was raised an obelisk in honour of the great Riquet, who cut a most intelligent trench through that isthmus with the same tenacity as de Lesseps two centuries later.

The obstacles

De Lesseps wrote to his mother-in-law, whom he had entrusted with contacting the Empress: 'My ambition, I fully admit, is to be in sole command of all the threads of this great affair... In a word, I do not wish to accept the conditions of anyone else, my aim is to have my own way in everything'. It is a letter that is highly revealing of a state of mind that won him the greatest glory in Suez but had dire consequences later.

In Constantinople, he was received extremely coldly and, in 1859, the Sublime Porte even ordered him to suspend all the work on the site. De Lesseps paid no attention and carried on regardless. In London, Lord Palmerston continued to affirm that the canal was a swindle but De Lesseps pleaded his cause tirelessly to British public opinion.

The financial obstacles would be overcome but apparently in less honourable conditions: in 1858, two hundred thousand shares worth two hundred million gold francs were offered to 'the capitalists of all nations'. The French subscribed a little over half the sum but in other countries the launch was a complete failure. It was then that Egypt agreed, at the urgent request of de Lesseps, to take up the 40% of remaining shares and had to borrow eighty-eight million francs. The *Compagnie Universelle* of the treaty had been reduced to a Franco-Egyptian Company.

---

24. For a history of that canal, see Jeanne Hugon de Scoeux, *Le chemin qui marche*, Loubatières 1994.

Egyptian historians[25] have not failed to underline all this, and there is no need to look elsewhere for the sources of the present coldness of the Egyptian Government towards the builder of the canal. They rightly consider that de Lesseps, in violation of the spirit of the concession contract and because he had been unable to get the 'capitalists of all nations' to buy shares, had dragged the country, poor as it was at the time, to the verge of bankruptcy.

The digging of the canal: the question of forced labour

The first blow of a pickaxe was given in April 1859, four years after the reconnaissance of the route. When the project was being worked out on paper, the problem of labour had been discussed. Mehemet Ali had never given this matter a thought for his major public works; like all the Pharaohs he had used the traditional system of forced labour, the corvée. As the very conservative Linant de Bellefonds[26] wrote, 'The ideas of progress and civilization, because that is what it is generally agreed to call them, had not yet led people to think of abolishing the corvée'.

Saïd wanted to be more liberal. He had forbidden slavery in 1854 and in 1856 had ordered the freeing of all those still in chains. Forced labour in the form of corvées would be a thing of the past and men would be 'supplied by the government in response to requests made by the engineers and as needed', with wages a third higher than the average daily rate in 1856 and payment of the cost of transporting the workers and their families. Despite these conditions, the takers were too few because the workers lacked the money to buy their food owing to the steadily falling value of the piastre. The old barter system of the Pharaohs and Mehemet Ali had its advantages! There were numerous desertions in 1860 and Saïd was then forced to fetch the peasants *manu militari* from all the provinces of Egypt: it was the return to the corvée.

The Viceroy died at the end of 1862, and his successor Ismaïl, at the request of the Sublime Porte, asked for the abolition of the corvée. De Lesseps rightly pleaded[27] that his enterprise was less deadly than the building of the British railways. Nevertheless, a complaint was lodged in Paris. The Emperor, asked to arbitrate, invited the Company to give up the requisition of workers in return for the payment of an indemnity of eighty-four million francs by the Egyptian Government.

---

25. See in particular Mohammed Sabry, *L'Empire égyptien sous Ismaïl et l'ingérence anglo-française*, Paris, Geuthner 1933.
26. Linant de Bellefonds, *op. cit.*, pp. 37-8 and 43.
27. The Company had created a workers medical service which reported 20 deaths out of 1250 European employees and 23 deaths out of an 'Arab population' of 120 933 persons between March 1861 and March 1862. In other words, as Robert Solé mischievously remarks, the canal killed proportionately one hundred times fewer Egyptians than foreigners. In the figures for Egyptian victims, was account taken of those who, exhausted, returned home to their distant villages to die? That said, we have here one of the first examples of industrial medicine.

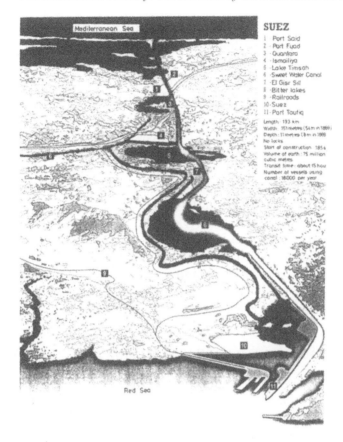

Figure 69. The canal built by de Lesseps, from B. Heimermann, *Suez et Panama*, Arthaud 1996.

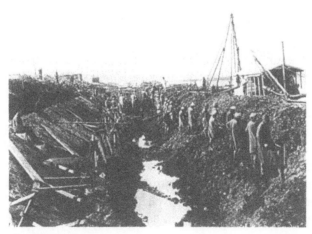

Figure 70. The corvée before the engineer Lavalley's invention of his big dredging machine (photo by Hippolyte Arnoux, from the archives of the Society in memory of Ferdinand de Lesseps and the Suez Canal).

De Lesseps had no choice other than mechanization. He was saved by the creativeness of the engineer Alexandre Lavalley, who designed for him the huge universally admired steam dredgers. De Lesseps was a first-class promoter[28] but, despite claims to the contrary, he had none of the qualities needed by a project manager. In spite of a cholera epidemic, the job was finished in 1868, just before an inauguration ceremony of unprecedented pomp.

No praise was then too fulsome for the creator of the canal. Tribute was rightly paid to his determination, to his tireless energy, his charisma and his honesty. Where many others had found opportunities for speculation and personal enrichment, he had remained perfectly disinterested.

Only Nubar Pasha, Ismaïl's Minister for Foreign Affairs, dared to point out that Egypt was about to be ruined because de Lesseps had not respected the spirit of the concession granted to him in a personal capacity by the all too trusting Saïd. While the world had been expected to offer a canal to Egypt, that country had had to pay through the nose: in addition to the eighty-eight million francs of forced subscription, there were the eighty-four million arising from the French Emperor's arbitration and soon another thirty to pay for various installations.

ISMAÏL THE MAGNIFICENT RUINS HIS COUNTRY

Opposed to the canal project from the start, Ismaïl quickly fell into the toils of de Lesseps. The naming after him of the new city founded on the shores of Lake Timsah flattered his self-esteem. In response to the pomp of the ceremonies organized in his honour at the Universal Exhibition of 1867, he played the munificent prince, handing out invitations and hosting banquets without sparing any expense. While in France, he had been able to observe the full extent of that country's Egyptomania and he decided, on his triumphant return to Egypt, to encourage a certain Francomania. The life of the royal household and of his subjects, as well as local fashions and the physical environment, would be profoundly marked by this.

The debts mounted up and in 1875 Ismaïl, after sixteen years of extravagance, was on the verge of bankruptcy: the impoverishment of the fellahin, who were forced to work with whips, now made it impossible to cover the costs of the administration. Ismaïl was forced to sell the shares in the canal at a time when traffic was expanding rapidly. The British, who had been so opposed to the canal, were lying in wait for their prey, and Disraeli bought up the block of Egyptian shares at a price that was hardly more than the issue price twenty years before. Shortly afterwards, ruined Egypt had to sell its right to 15% of the profits. The country was far from the great hopes awakened in Saïd during his first discussion with de Lesseps. Ismaïl the magnificent had thought it possible to live off the in-

---

28. Nathalie Montel comes to the same conclusion in her book *Le Chantier du Canal de Suez*, Presses des Ponts et Chaussées, Paris 1998.

come from the shares in the canal and had seen himself following in the footsteps of Amenophis III, the Pharaoh of Egypt's golden age. On 26 June 1879, he received an unexpected telegram from Constantinople: ironically addressed to ' the former Khedive of Egypt', it told him he had been deposed.

ANOTHER PHARAONIC DREAM – WITH DIRE CONSEQUENCES FOR FRANCE

Ismaïl was the first victim of *fir'aùnia*, de Lesseps the second. Those who sojourn in the Nile valley, that muse of great designs, do not always escape lightly. De Lesseps had been untouched by the epidemic of plague but he was caught by a virus that had been fatal for the last viceroy. A fine figure of a man and a brilliant horseman, he was much admired and the great monuments scattered over the Egyptian plain – pyramids, temples and other works of man – seemed to him filled with the same grandeur as the canal he had just carved into the land. This feeling would nurture in him a desire to do even better.

The virus developed at the moment of the inauguration: no words were rich enough to sing his praises. The score of *Aida* includes an undoubted allusion to him when, in Act II Scene 1, a member of the court of Ramesses III declaims:

O hero who, to the sound of hymns and prayers,
flies off to glory.

He was laden with honours and decorations. It was just after the terrible military defeat of the French at Sedan in 1870 and the subsequent occupation, and France was searching for someone to admire. He was called 'the great Frenchman'. The Academy of Science opened its doors to him in 1873 even though he was far from being a scientist; he was co-opted into the *Académie Française* even though his very careless style of writing did not appear to fit him for approval by a company of literati.

He was attracted by great projects: after a visit to the shotts – saline lakes – in Algeria and Tunisia, he expressed himself very much in favour of turning them into an inland sea whose waters would be fed by the Mediterranean through a canal. Plans were drawn up by Commander Roudaire, but some engineers soon discovered that the areas to be flooded were above the level of the Mediterranean!

Then another isthmus started to haunt him: was he not called the piercer of isthmuses? Someone mentioned the isthmus of Corinth to him, but it was too short to interest him. His eyes were not on shortening the journey round the Peloponnesus but on joining two great oceans, the Atlantic and the Pacific.

Thus began the adventure of Panama. History has been excessively indulgent in continuing to see him as the innocent and unconscious victim of unscrupulous financiers and crooked backers. His most recent biographer[29] goes so far as to offer him as an example for French youth and to speak of unjust judges at his trial. As André Siegfried[30] has rightly emphasized, Panama was also a story of the building of a canal and that history, in my view, shows clearly how personal hubris and the desire to surpass others can lead a great man to stoop to sinister manoeuvres at the expense of his integrity. The consequences were immense: some twelve thousand deaths and losses of a billion and a half francs, three times the cost of the Suez Canal, which ruined thousands of small French investors.

At Suez, de Lesseps, because he had been surrounded by clever engineers, had created a false impression of supreme competence. And yet he had had mixed views about those engineers, considering that they took delight in complicating matters: he felt he would do better to rely on his instinct and common sense. In other words, his experience at Suez led him to reason by analogy, which was taking a fearsome risk in fields such as climatology, epidemic medicine, hydraulic engineering and the earth sciences.

He waved away all objections with 'I did Suez and I know what it is all about'. He simply followed his own inclinations: he had decided that the Panama canal, like its twin brother, would be at sea level without locks. If it was objected that there was a difference of 6 m between the spring tides of the Pacific and Atlantic oceans, he would reply that, like Lepère, geographers made mistakes and that locks were unnecessary, as unnecessary as the locks imagined for Suez by Talabot the Saint-Simonist. What about that tumultuous river, the Rio Chagres, which crossed the path of the canal? There existed nothing like it in the Suez isthmus. What about the Culebra Cordillera, a ridge over 100 m high? He would answer by pointing to his success in getting the canal through the Gisr Sill with its maximum height of 17.50 m. So what, if the rotten soil of the Culebra Cut became a soft paste under torrential rain. Did people really believe that he knew nothing at all about the defects of the soil when at Suez he had rapidly had his way with an unexpected bed of rock? So what, if a murderous climate had swallowed up thousands of human lives in the construction of the interoceanic Panamanian railway? Did people really believe that the digging of the Suez canal under a boiling sun had been a picnic?

In vain was it explained to him that one must adapt to nature, feel one's way carefully without forcing it. In vain too did Godin de Lépinay, a truly great engineer who had worked in Central America, speak out about the unhealthy climate and the defects of the project and put forward a rational and straightforward counter-project with locks that the Americans[31] would subsequently adopt in the

---

29. Ghislain de Diesbach, *Ferdinand de Lesseps*, Perrin, 1998, pp. 17 and 403-423.
30. André Siegfried, *Suez and Panama*, New York, Harcourt Brace 1940.
31. It has been mistakenly asserted that the Americans took up the project of de Lesseps; they always considered his project impossible to carry out. Philippe Bunau-Varilla, whom they called the Bonaparte of engineers, saved the honour of France by getting them to adopt, some twenty years later, the project of Godin de Lépinay.

first decade of the twentieth century. The great Gustave Eiffel seconded all these criticisms during the impassioned discussions of an international commission meeting in Paris. Most of its members were from other countries, had unbounded admiration for de Lesseps, had been selected by him and had never set foot in Central America. They disregarded the critics. Before the vote, Godin de Lépinay made a solemn statement: 'In order not to burden my conscience with these unnecessary deaths and with the loss of a considerable capital... I shall vote No'. Prophetic warning!

De Lesseps now became guilty of serious disinformation that bordered on breach of trust. He sent the *Académie des Sciences* a few samples of a hard rock which had only a distant connection with the soils of the Culebra, which became like cakes of soap under tropical downpours. The illustrious Academy then certified[32] that the deep trench of the Culebra Cut could be executed by their fellow-member without difficulty and with almost vertical sides, rather like the Corinthian Canal. De Lesseps would take the floor in the great schools that had provided him with so many skilled engineers to say what an exalting task awaited the volunteers and how healthy was the country of their future exploits (he himself went to Panama three times, but always in the dry season when there was no yellow fever). So many engineers and workers lost their lives there that the Americans called de Lesseps 'the great undertaker'.[33] As Petit, a graduate of the *Ecole Centrale des Arts et Manufactures,* is recorded as saying to Sordoillet, a graduate of the *Ecole Polytechnique,* 'We have arrived together and we will die together'. And events soon proved him right. And yet, in spite of such disasters and even when the project was at its worst, the communiqués to shareholders and future investors were always full of confidence.

De Lesseps, a stubborn old man with his halo of glory, had absolutely no sense of moderation and no critical spirit. Speaking of such excessiveness, which he called *hubris,* Aeschylus wrote: 'It is an evil with which everyone is familiar for having cursed or committed it. It is a thought of pride based on idolatry or on self-worship. The only way to escape nemesis, or divine vengeance, is not to get caught in its toils'. De Lesseps did not escape the anger of the despoiled investors. Already in poor health, he was sentenced to five years imprisonment; so was his son, Charles, who had nevertheless tried to warn his father since the very beginning; Gustave Eiffel, too, whom he had finally succeeded in carrying with him, was given two years. 'Beautiful illusions', conceded his defending counsel to the Court at his trial. For me, the only mitigating circumstance that can be offered is that de Lesseps was branded with the same ambition to leave his name to poster-

---

32. Minutes of the *Académie des Sciences,* 1880, p. 275.
33. See D. McCullough, *The Path between the Seas: the Creation of the Panama Canal (1870-1914),* New York, Simon & Schuster 1977. McCullough gives an excellent account of the second part of de Lesseps's life, from 1870.

ity as the great Pharaohs, and that this sometimes leads to irrational behaviour. There was something in de Lesseps that reminds us of Cheops adding slab after slab to roof the burial chamber in his pyramid or of Ramesses II ordering the construction of his tomb on an impossible site. But it is hard to imagine such a historical argument securing the clemency of the judges!

France was made poorer, Egypt was ruined, and the dying de Lesseps had lost his honour and sullied his Egyptian glory, while the steadily increasing traffic through the Suez Canal showed how right he had been to inscribe that blue ribbon of Saint-Simonist dreams in the soil of Egypt.

What a curious destiny awaited those who cut through isthmuses. Like de Lesseps,[34] Riquet ruined himself in building his canal, but he at least did not have

Figure 71. The statue of Ferdinand de Lesseps, by the sculptor Emmanuel Frémiet (1824-1910), used to look out over the entrance to the canal at Port Saïd. It was 7 m high and placed on a pedestal of 10 m. In 1956, it was blown up. Recovered and repaired, it is now stored in a hangar awaiting another destiny... (Photo from the archives of the Society in memory of Ferdinand de Lesseps and the Suez Canal).

---

34. On 5 June 1894, the general assembly of shareholders in the Suez Canal Company voted Madame de Lesseps and her children a life pension of 120 000 francs to assure their future. As for Riquet, ruined after having invested his entire fortune in his canal, he wrote to Colbert, the Finance Minister of Louis XIV in around 1670: '*I regard this work as the dearest of my children; this is so true that, having two daughters to settle, I prefer to keep them with me a bit longer and to use the money I had set aside for their dowries to pay for my canal*'.

deaths on his conscience and had not resorted to loans and people's savings. After a glorious start, Riquet's canal was used less and less – quite the opposite of the Suez Canal. Tribute would be paid to de Lesseps five years after his death, when the British, who had been his fierce opponents, agreed with the French to erect on the banks of the canal a statue of the person who had finally surpassed the works of Necos, Darius and Trajan.

THE SUEZ CANAL, A FESTERING WOUND FOR THE EGYPTIANS

After the inauguration of the canal, Egypt was bankrupt and owed its survival to the forbearance of its creditors. But it would only be a short breathing space: the Khedive, strangled by debt, was forced to sell his shares to the British banks.

In 1882, the British bombarded Alexandria and their army occupied Egypt on the pretext of protecting the foreign powers from the ambitions of an officer of Egyptian origin, General Orabi, who had dared to speak out against their seizure of the country. For Egyptian nationalists, the canal built by virtual slaves had become a festering sore: the Suez Canal Company was a state within the state and the canal, after having ruined Egypt and made the fortunes of many British and French, would become the target of demands by the Wafd, an independence movement led by Zaghlul Pasha.

In 1882, a revolt fomented in Sudan by Mohammed Ahmed, known as the Mahdi, would serve as a pretext for Britain to put an end to the Franco-British condominion and take all power for itself. In 1915, in a final convulsion of the Ottoman Empire, the Turks attempted to reconquer the canal through which the British troop reinforcements arrived, but they were repulsed.

CHAPTER 4

Egypt in modern times

The British protectorate became a sovereign state and Egypt was granted nominal independence in 1922, after three years of violent disturbances. The Sultan Fuad had himself addressed as 'His Majesty the King of Egypt'. The kingdom lasted only thirty years. In 1951, nationalist resistants harried the British troops who continued to be stationed along the canal and, in January 1952, the latter massacred a company of Egyptian police suspected of hiding commandos. Riots broke out in Cairo and buildings were set on fire.

NASSER, AHMOSE OF MODERN TIMES, PHARAOH OF LIBERATION AND ANOTHER CASE OF HUBRIS

In 1952, the 'Committee of Free Officers' took power and Farouk, the last king, left the Summer Palace in Alexandria for exile. Two years later Nasser became the Rais, the 'boss' of Egypt.

Nasser dreamed of constructing an enormous dam that would protect the country for ever from the devastating consequences of floods like those suffered by the Pharaohs. After having been refused by Washington, London and Paris, he was supplied with arms by Czechoslovakia. Despite this and despite also the opinions of intellectual groups in Cairo, he decided to renew his appeal to the West regarding the dam. On 19 July 1956,[1] John Foster Dulles, the American Secretary of State, informed him that 'in view of the instability of the Egyptian economy' the loan that had been promised the previous January would not be accorded. The Egyptian people viewed this as an insult and, in front of a delirious crowd, Nasser proclaimed 'I take the canal', confirming what Orabi had said seventy years before: 'The canal is in the service of Egypt and not the other way round'.

The war that followed is called 'The Suez War' by the British, the French and the Israelis, but in Egypt it was regarded as 'a cowardly triple aggression'. The statue of de Lesseps was thrown into the sea.

---

1. See Appendix 2, 'The High Dam, a challenge to the West'.

The occupant was kicked out just as another occupant had been in around 1650 BC by Ahmose. The Prince of Thebes, having become Pharaoh, had expelled the Hyksos (literally 'foreign princes') from Syria, Palestine and Sinai. Egyptian nationalism was exacerbated and the memories of long-past invaders reawakened – Bonaparte, the shelling of Cairo, the Battle of Imbaba. They even went so far as to celebrate the victory of Saladin over the crusaders and to commemorate the defeat of Saint Louis in 1249.

It was also the beginning of an intense *fir'aùnia:* patriotic films such as *Rends-moi mon coeur* sang the revolution; the television studios busied themselves with productions and sets based on the great civilization of the distant past.

In Egyptian politics and literature, admiration for the spirit and achievements of the Pharaohs lay just under the surface. The novelist Naguib Mahfouz, a recent Nobel prizewinner for literature, devoted three novels to the Egypt of the Pharaohs and school textbooks emphasized the historical primacy of Egypt. The Pharaohs were idealized and endowed with all the virtues; the man in the street, hearing about financial scandals, would wax indignant and exclaim 'Are we really the heirs of the Pharaohs?' The Pharaohs had rejected their own ancestry, but the Egyptians like to see themselves as descendants.

Even sport was not spared: people were convinced that the Pharaohs were all-powerful. The players in the national football team had the right to call themselves 'the Pharaohs' after their victory over Zambia in 1998 and a journalist, in celebrating the merits of the victors, wrote: 'The two playmakers of the Pharaohs are able at any moment, by a stroke of genius or a pass of the utmost precision, to change the course of the match'. The same reference to the past in handball: 'Our Pharaohs are rapid and creative' or 'The young Pharaohs need experience'.

But, above all, this belief in the Pharaohs became an invitation to bold action on a grander scale than anything before. In foreign policy, Nasser dreamed of following in the steps of the most glorious Pharaohs and of pushing the frontiers of his country beyond the Red Sea. Already, in a somewhat disorderly agitation, he had embarked on adventures – without success – in Syria and Yemen. The spies of Ramesses II had misled their king: the Hittites had a bigger army than he had imagined. It was the same with Nasser when he conducted his war against Israel. Nasser was an admirer of the military genius of Napoleon but, like Napoleon after the siege of St Jean d'Acre, he had to retreat rapidly in the Six-Day War. On his return, however, he did at least have the modesty to take the blame – unlike Napoleon, who had hosted a celebratory dinner in Cairo.

Little by little, from the depths of history, the Egyptian distrust of alliances or close relations with countries to the East rose to the surface. 'I have killed the Pharaoh', would later confess one of the assassins of President Anwar Sadat, whom he saw as a Pharaoh of evil since the President had, one evening in November 1977, decided to go and shake the hand of an enemy. For, in Egyptian eyes, the Pharaoh embodied both evil and its opposite: probably owing to the influence of the Koran. Surat VII, for example, describes at length the meeting between Moses and the magicians of the Pharaohs. The latter thus became enemies of the true faith, that of the Hebrews in captivity and then of the Muslims. That is just one of the ambiguities in Egyptian attitudes to the pharaonic past.

When we turn to the major projects of Nasser, we see the memory of the grandiose architectural ambitions of the Pharaohs resurfacing, but now tinged with a religious influence. The Koran says: 'This is a people that have passed away;they shall have what they earned and you shall have what you earn, and you shall not be called upon to answer for what they did'. It was no longer a question of admiring the pyramids but of drawing inspiration from them. And this explains why, under Nasser, the country allowed itself be carried away by a wave of large-scale projects.

THE HIGH DAM: IGNORANCE OF THE WISDOM OF THE PHARAOHS

Long before the construction of the High Dam, irrigation schemes based on reservoirs created by dams had been established, first of all in the delta from 1840 and then at Asyut, Nag Hammadi and Aswan. These dams had been small in size but had rendered considerable services because they had facilitated the drainage of land downstream without recourse to pumping upstream. They made the rhythm of the seasons perceptible: in winter and in spring they stored the water and at the summer solstice spread over the plain, through numerous sluices, the silt-laden inundation that brought fertility.

The question now was to build, at Aswan, upstream of a dam constructed in 1902 by the British, a very high dam that would become as powerful a symbol as the Suez Canal had become. The most recent raising of the British dam, which submerged the Temple of Philae during each inundation, had shown that the real solution was not to build higher dams but to build more of them. In fact, without prospecting for other sites upriver, Egypt remained a prisoner of the sacred site 'between Crophi and Mophi', the two mountains where, as we have already learned, the very ancient Egyptians situated the sources of the river, *in the fathomless depths*.

After the slap in the face by Foster Dulles, the Raïs had therefore announced that he was nationalizing the canal, whose revenues would be used to finance a large dam. In 1958, the Soviet Union granted additional financing and assistance with equipment and manpower: Egypt had its revenge. Then began the construction of that enormous, still controversial dam.

The final plans were drawn up in Moscow, at the *Hydroprojekt Institute*. At the time, Soviet architecture was marked by the post-war megalomania and found itself completely in tune with the pharaonic ambitions of Nasser. The result was a mountain of rock and sand laid across the river, nearly a kilometre thick at the base, a 110 m high and seventeen times the volume of the Great Pyramid. This 'cold monster' had what it takes to impress the masses, who saw it as a striking proof of their liberation from a hated imperialism and a sign of Egypt's leadership of a new world, the Third World, that was fighting for its independence. To sign their work the Russians erected, on the left bank near the High Dam, a metal sculpture in the form of a stylized flower that contrasted sharply with the beauty and simplicity of the temples of the Pharaohs.

Behind the dam the oncoming waters were to swell the Nile for some 500 km upstream, including 150 in Sudan. The first two cataracts would disappear and a large proportion of Nubia would be swallowed up under a gigantic reservoir of 168 billion m^3 of water, making Lake Nasser nine times the size of the Lake of Geneva.

By 1960 'the imperialists', as the Egyptians liked to call them, had realized that the Egyptian determination was irreversible: with a spontaneity as rapid as their earlier refusal to fund the scheme, the Americans participated so generously in the rescue operation that the Egyptians gave them the temple of Dendur, the first stones of which arrived at the Metropolitan Art Museum of New York in 1967, only eleven years after the icy rebuff of Foster Dulles. In the end, under the auspices of Unesco, about fifty other nations, led by France and Italy, came to the rescue of seventeen Nubian temples doomed to be engulfed. A most fortunate reconciliation that made possible the saving of an exceptional world heritage. It is impossible to praise too highly the successive directors of Unesco, Vittorino Veronese and René Maheu, and the Frenchwoman Christiane Desroches Noblecourt. Unesco would do much better nowadays: when an important element of the world heritage is in danger, it visits the site with a group of experts and issues a strongly reasoned opinion which often leads to the abandonment of the destructive project. This happened in 1995 when, on the advice of a commission chaired by Mounir Bouchenaki, the Portuguese Government halted the construction of a dam on the Coa river that would have submerged a site exceptionally rich in rock carvings about twenty thousand years old. The gravity of the decision can be measured by the fact that Portugal has to import nearly 90% of its energy.

A SLOWLY GROWING AWARENESS: THE HUGE SPILLWAY LIKE A CANNON TRAINED ON THE BUSY PLAIN

The two main protagonists, Nasser and Brezhnev, died well before the reservoir was full. But as the water rose, people began to wake up. There were probably some misgivings about the drowning of an entire community but there was not all that much concern about the past: they were rather wondering whether the huge project fitted in with the country's history and geography.

After all, was that man Brezhnev, who in his verbal sallies did not hesitate to compare Nasser with Hitler, really fond of Egypt? Had he really taken all the sides of the problem into account? Was not the most important thing for him, in his opposition to the United States, to strengthen the image of a new Third World allied to communism?

Moscow was such a long way from Aswan. Anyone drawing up plans for such a large project should first explore the past, as the architects of the Renaissance had done. In this case, it was the duty of the planners to understand the history of the river, to feel its pulse over the centuries, to study the shaping of its banks and, most important of all, to make themselves familiar with the knowledge possessed by the Pharaohs.

The Soviet scheme was a general-purpose project. On closer inspection, it resembled a dam, the Sadd el-Kafara, built by a Pharaoh on the Wadi Garawi and now regarded as the oldest dam in the world.[2] This dam was designed to check the floodwaters of this right-bank affluent of the Nile a little upstream of Memphis. It was constructed around 2500-2600 BC, that is to say four centuries after Narmer and forty-five centuries before Nasser. It was remarkably intelligent in design, with a very wide dyke like that of the High Dam and, again like the High Dam, with an impervious core containing a large proportion of clay to combat infiltration. Progress is not always what it seems.

It is true that the Sadd el-Kafara dam met with an accident: an exceptional flood occurred just before it was completed, filling the reservoir beyond its capacity; the dam was submerged and its central section destroyed. This is a danger lying in wait for all earthen dams, including the High Dam, since all such dams are constructed of loose – and hence erodable – materials, in many cases in ignorance of the amplitude of the most violent floods. Below Sadd el-Kafara, however, there were no homes or cultivated land whereas in the valley below the High Dam live today some sixty-three million people.

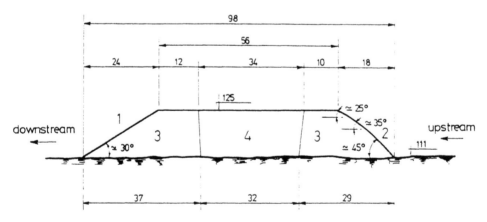

Figure 72. Cross-section of the Sadd el-Kafara dam, built by a Pharaoh towards 2500 BC on the Wadi Garawi (from Garbrecht). 1 = downstream slope; 2 = upstream slope; 3 = rockfill; 4 = impervious core; dimensions in metres.

2. G. Garbrecht, 'Sadd el-Kafara: the world's oldest large dam', in *Water Power and Dam Construction*, July 1985.

Figure 73. Ruins of the Sadd el-Kafara dam.

Figure 74. The High Dam (with the upstream face to the right) and the head of its creator: a document of Nasserian propaganda.

And so, to prevent the High Dam from being swept away by a giant flood, the Russians inserted into the dam, a little below the level of its crest, a spillway reputed to be non-erodable, a giant overflow ready to vomit exceptional floodwaters into the valley. When all is said, dams are an expression of man's powerlessness: 'I have barred your course but it is not in my power to contain your great fits of anger!' And that is the real danger of gigantism, for very great fits of anger can be very difficult to subdue. Indeed, was there not a contradiction between the desire to encourage urban and rural development right up to the foot of such a huge dam and the presence of that enormous spout, trained like a water-cannon on the fields and villages?

It had been explained to Nasser that the Nile of the twentieth century no longer had anything in common with the Nile at the time of the Pharaohs. Between 1840 and 1770 BC there had been a number of phenomenal floods, three times as strong as those of our nineteenth and twentieth centuries. Since then the Nile had become much calmer, with a fairly regular variation in flow that was slowly decreasing with time (Fig. 75). Nasser was assured that a less gigantic reservoir would easily absorb the biggest inundations of the last hundred years. The young dictator replied that one must beware of statistics: it was true that, thanks to the nilometers, the scribes of the Pharaohs had measured the heights of the floodwaters for centuries but there was much speculation about the units used by those instruments and about the exact value of the cubit, the unit of measurement used for heights. And he added that rivers are like volcanoes, with sudden rages. If El Niño had been known at the time, he would have taken it as an example. Nasser the liberator was dreaming of an immense dam that would slake the thirst of the Egyp-

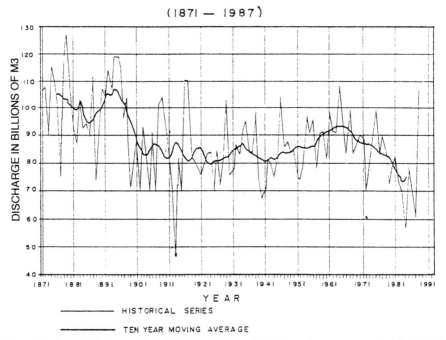

Figure 75. Annual discharge of the Nile from 1871 to 1987 (Egyptian source, quoted in the 1993 minutes (p. 79) of the International Commission on Large Dams at Cairo (I.C.O.L.D.)). These records would justify the annual discharge of 84 billion m³ adopted for the water-sharing agreement with Sudan. Note that the annual discharge appears to be decreasing with time.

tians and their fields for eternity. His advisers, to justify the hubris of their master, drew attention to 'erroneous' data, asserting that in 1911-1912 the annual dis-charge had dropped to 12 billion m^3 of water,[3] even though it is acknowledged (Fig. 75) that it in fact descended to only 48 billion. This error would have serious consequences. He was also assured that the centennial floodwaters had attained 155 billion m^3 in 1878 whereas they had in fact attained only 128. In the end, it was decided to increase the capacity of the Egyptian reservoir from 130 to 168 billion m^3![4]

THE CONSEQUENCE OF HUBRIS: SUCH A LONG TIME TO FILL THE RESERVOIR

Those trying to save the temples had been told that the waters would rise rapidly and might reach their maximum level shortly before 1970. That was the year in which Nasser died, but behind the dam there were scarcely 40 billion m^3 of water rather than the 168 billion that the reservoir could hold. Could it be that the Raïs, before he died, asked himself whether he had seen too big? For that is what he had done: the Nile will never totally fill the reservoir. In 1978 its volume was still only 134 billion m^3; it fell to 40 billion in 1987 and eventually reached 155 billion in 1998, and a little more in 1999 some 34 years after the damming of the river.

After the death of Nasser, Sadat re-examined the matter. It seemed to him to be dangerous to have this high dam overhanging a plain in which all the wealth of the country was concentrated. It should be added, moreover, that Sadat, strongly anti-communist, was always ready to gainsay his predecessor. Nor did he have total confidence in Soviet technology: he was perfectly aware that dam spillways, owing to the violence of the water pouring over them, sometimes eat into the base of the dam they are meant to protect, like birds pecking away at the branch on which they are perched. And so, when the solidity of the Soviet overflow system was tested, its weakness became apparent: under the impact of a minor flood, half the size of those which the smaller dams further downstream could let through without difficulty, it showed signs of deterioration. Sadat gave orders to look for another way of evacuating the floodwaters.

Sadat's successor would support this approach. The immense reservoir terrified President Mubarak at a time when his neighbour, the fiery Colonel Gaddafi, threatened to bomb the dam: if the immense mass of water it contained were to be released, experts said that the whole of Egypt would be drowned. Relations with Libya have since improved but, at a time when atomic weapons are no longer the prerogative of the big powers and potential conflicts between Egypt and the countries of the Nile's upper reaches are in the offing, it remains a burning issue.

---

3. This false figure is still bandied about today by unconditional admirers of the High Dam.
4. See the general conclusion of the analytical report on the Egyptian project by the *International Bank of Reconstruction and Development*, dated 28 February 1956.

Figure 76. Lake Nasser covers the first two cataracts and about twenty temples.

Boutros Boutros-Ghali,[5] then Sadat's Minister for Foreign Affairs, concluded after his 1978 African tour of the 'depths of the Third World' that 'our African continent suffers from economic backwardness, but it is infected with something more dangerous, and this is the power-mad delusions of some of its rulers. We cannot achieve development in Africa unless we succeed in building the African individual. And we cannot begin to build the African individual until power-mad despots like Idi Amin and the emperor Bokassa disappear from the scene.'

Even when this *homo africanus* is forged, the High Dam will still be well within the 1500 km range of Israeli missiles. The 1992 Amsterdam crash of the El Al plane carrying the gas Sarin has made it clear that, as a last resort, Israel was ready to employ means dangerous for its enemies.

---

5. Boutros Boutros-Ghali, *Egypt's Road to Jerusalem: a Diplomat's story of the Struggle for Peace in the Middle East,* p. 99.

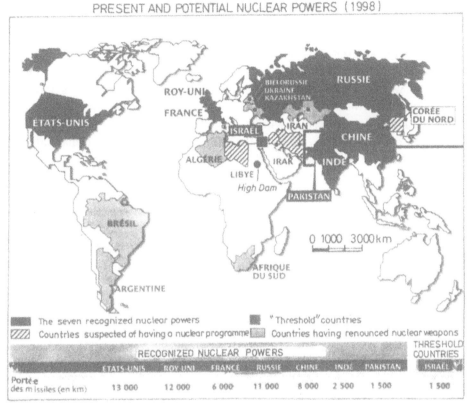

Figure 77. The High Dam and the nuclear powers.

In fact, one of the lessons to be drawn from the NATO-directed war for the liberation of Kosovo is that blunders in aerial bombardment can be numerous: a hospital can be mistaken for a power station. But a missile guided by two computers, one reading the topography of the land and the other an image processor, can identify a target to within a metre and could not miss a high dam, which will thus become a favourite target in the wars of tomorrow. The Aswan Dam and its reservoir, the largest in the world,[6] overlooking a densely populated plain which also contains a large proportion of the world's heritage, has become a matter of grave concern.

The large reservoir desired by Nasser as both a remedy and an emblem is a real danger. It seems that the Raïs, sensing future conflicts over the sharing of water, had wished to lay in provisions, like a housewife when a conflict looms, and, to quote the French humourist Pierre Dac, 'To set on one side in order to have

6. With the exception of the Kariba Dam on the Zambezi River, a special case owing to the gorge through which the river flows.

enough in front remains the chief concern of all those who have grasped the need to get ahead in order to protect their rear'. The flow from the tap had been confused with the size of the basin. The project turned out, as it were, to be a hydro-political scheme and not a rational example of hydraulic engineering. It thus becomes of interest to find out about the size of the reservoir planned for the largest dam in the world, that of the Three Gorges across the Yangtse-kiang River in China. That dam will attain a height of *180 m* – almost double that of Aswan – but the capacity of the reservoir will be only *39 billion m³*, less than a quarter of the capacity of Lake Nasser.[7] And yet the valley of the Chinese river has suffered in the past from exceptionally deadly floods, with 145,000 victims in 1931, 142,000 in 1935 and a lot more than that in 1870, not to mention the floods of 1998 still fresh in the memory.

The outsize nature of the Aswan High Dam, and the enormous wastage of water, became obvious: average annual evaporation from the surface of the lake amounts to some 10 billion m³, a major handicap when water is becoming an increasingly precious commodity. What had happened to the goddess Ma'at, who used to symbolize reason and moderation in the valley of Egypt?

The Chinese Three Gorges Dam calls for further reflection. The go-ahead was given by Li Peng, Chairman of the National People's Congress, who had trained as a hydrologist. The construction work was already well advanced when it was halted for a time, on the pretext of a momentary shortage of funds, by his successor Zhu Rongji, who seemed to have about as much sympathy for his predecessor as Sadat had had for Nasser, but it is also possible that the stoppage masked some further reflection about the vulnerability of major works of hydraulic engineering in case of war.

A RETURN TO THE WISDOM OF THE PHARAOHS: ADMISSION OF THE ERROR IN THE SOVIET PROJECT

It will be remembered that the ancient Pharaohs, in order to attenuate the force of the river in flood and nourish with silt the fields all along the valley, made cuttings through the embankments. In their time the Nile was, as we have already said, like a pelican nourishing its young with its own flesh. Gigantism was ruled out. At a minimum, if the Soviets had been willing to take an interest in the history of ancient Egypt, they could not have failed to be aware of the existence of Lake Moeris, a kind of depression beside the valley where the Egyptians used to store part of the excess water from the most powerful floods. If they had been interested in the history of Babylon, they could have read the words of Arrian recounting the history of the Pallacopas: 'The Pallacopas lies at 800 stadia from Babylon... As the current of the Euphrates carries a considerable volume of water

---

7. When the Boulder Dam – the largest in the world at the time and now called the Hoover Dam – was at the planning stage, its reservoir was designed for a capacity of only 40 billion cubic metres.

in spring and summer, it bursts its banks and spreads over the land of the Assyrians... It would flood the country if the people did not make a breach in its banks to let the water flow into the Pallacopas and be diverted towards the marshes and lakes that begin at this canal' (Anabasis, VII, 21, 1-6).

The ancients saw the Nile as a 'benediction and doom'. The 'doom' was the occurrence of truly exceptional floods. It was absolutely vital to bring this threat under control as far upstream as possible, at the frontier of the country.

A few treatises on geology and geography were consulted and it was discovered that the uplifting of Nubia had shifted the bed of the river westwards so that nowadays the Nile, flowing along the edge of the Libyan desert, dominated a very ancient nearby depression in the desert containing the Neolithic site of Tushka and, further to the east, the depression of Nabta Playa where forgotten peoples of the desert had developed an astonishing civilization. The area is bordered to the north by a large limestone plateau (Fig. 78) at the foot of which lie the oases of Kurkhur and Dunqul, today uninhabited. Here geology and ancient history had come together: opposite this site the Nile sweeps round a large rock and in so doing makes such a peculiar sound that a British map calls it 'the rattle of Tushka'. Chephren, of whom around a hundred statues have been found in the area, had marked with a stela the quarry from which he extracted his favourite stone, a

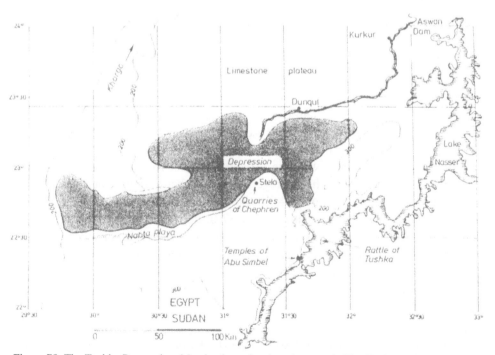

Figure 78. The Tushka Depression. Map by the author based on one in *The Geology of Egypt*, by R. Saïd, pp. 10-14, satellite maps and an Egyptian document presented to the I.C.O.L.D. at Cairo in 1993 (Minutes, p.253).

plutonic rock resembling diorite. Indeed, any atlas using colours to indicate relief – such as Figure 15 – shows the existence of a deep depression in the neighbourhood of the Tropic of Cancer from 30° to 32° East.

It was then remarked that a Sudanese counter-project which would have saved a good part of Nubia had been rejected too hastily. It had advocated a cutting in the left bank at Batn-el-Haggar,[8] level with the second cataract, through which excess floodwaters could be directed towards the Tushka Depression. It was an interesting idea, but rather ambitious, since Batn-el-Haggar was quite a long way south of Tushka.

In the end, after more than twenty years of reflection, the idea of a lateral spillway was taken up, flood waters being diverted into the depression. The construction work was carried out between 1978 and 1982 without much publicity because it in fact amounted to condemnation of the Soviet project. It was a lesson in the proper use of history and geography before embarking on a major project.

The Tushka Depression has sufficient capacity[9] to receive the total discharge of the Nile for two years! In 1998, following unprecedented floods in the tropical belt, the reservoir at last reached the level of 180 m, still 3 m below the maximum figure allowed. 'Emergency at the High Dam' ran the headlines in the press and those loyal to the Nasser era consoled themselves with the rightness of the judgment of the Raïs, even though the diversion of water to Tushka had made all panic unnecessary. The Egyptian Ministry for Hydraulic Resources quietly admitted that the Tushka site had received only five per cent of its capacity.

A SMALLER DAM ON THE EGYPTIAN BORDER WOULD HAVE SUFFICED

It was now obvious that the dam could have been built close to this natural spillway,[10] that is about 250 km further upstream, and have been designed for a capacity just sufficient to regulate the summer discharge.

When we take another look at the official Egyptian figures published in 1993 (see Fig. 75), we find that the average annual discharge over a period of more than a hundred years was around 85 billion m^3. From 1871 to 1987 it fell only once to 48 billion (in 1912), that is to say 37 billion m^3 less than the average. If we allow for the possibility of such a drought occurring in two consecutive years, for evaporation and for the deposition of silt, it emerges that a reservoir of 80 to 90 billion m^3 – half the volume of Lake Nasser – would have sufficed.

This reservoir would have been what is now the southern half of Lake Nasser.

---

8. J. Vercoutter, *Les forteresses égyptiennes du Batn-el-Haggar ou les failles d'une campagne.*
9. The depression has an area of roughly 6000 km^2 and its level ranges from 121 to 180 metres above sea level, with the lower part in the west. The maximum elevation of the Aswan reservoir is 183 metres.
10. There the Nile gorge is less narrow than at Aswan so the dam would have been wider but over 20 m lower, firstly because the land was about 20 m higher and secondly because the level of the reservoir could be maintained a bit lower.

There are six other dams on the way to the sea, three of them between Aswan and the delta. Indeed the Egyptians had wisely renovated some of these dams and slightly increased their capacity. Two or three other small dams of a similar type, which had been left unfinished in the general upheaval, could have been completed; Egypt would then have had total reserves of water of well over 100 billion m^3,[11] largely sufficient and much better distributed. We would then have had a Nile flowing down towards its mouth without a giant step. It would have represented a return to the ancient wisdom of the Pharaohs, with their small transversal dykes cascading in the plain. The Soviet scheme led to a break with the wise traditions of the past. It had confused grandeur and moderation.

UNNECESSARY DROWNING OF THE TEMPLES AND VILLAGES OF LOWER NUBIA

The builders of the High Dam therefore bear a heavy responsibility not only towards the Nubians but also towards the international community which had so generously stirred itself to save the temples of Lower Nubia.

The Nubians have always been the sacrificial victims of the valley's history, all too often used as mercenaries. They gave everything and received nothing in return, bringing their gold and their gifts to the Pharaohs. Among the many different ethnic groups that live all along the banks of the Nile, their natural honesty, kindness and devotion make them particularly attractive. Today, some eighty thousand of them have been exiled into concrete shacks that have nothing at all in common with their homes nestled in the meanders of the Nile valley, whose architecture made the region so charming. They desperately regret the tragedy that has destroyed all their links with their past. Henri Froment-Meurice, the French Ambassador and acting chargé d'affaires in Egypt just after the Suez business, writes: 'As for me, the only chance I had to visit Nubia was thanks to a supply boat belonging to the Department of Antiquities. Although I was naturally fascinated by the beauty of the temples standing in the waters of the Nile, I was saddened just as much by the spectacle of boats filled with Nubians leaving their homes, adorably painted and decorated with glasswork, with their furniture and animals – and not all their animals because in those deserted villages doomed to be submerged roamed starving dogs howling at their approaching death'.[12]

In a film called *Un jour, le Nil* made at Aswan, the Egyptian film director Youssef Chahine rose up in revolt against the building of the High Dam that was going to wipe off the map the ancestral lands of the Nubians. His film was blocked by the censors as soon as it came out, in 1968.

---

11. The total capacity of all the existing dams above and below the High Dam comes to almost 50 billion cubic metres.
12. Henri Froment-Meurice, *Vu du quai*, 1998, pp. 260-261.

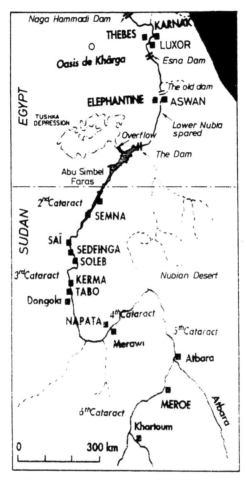

Figure 79. A more rational location for the High Dam, if it had been constructed after careful study and without political ulterior motives.

Old Nubians who had descended the Nile as far as Cairo learned that the great city to be protected was less than a thousand years old, whereas their country had a history dating back five thousand years. They could never understand why so many foreigners made such haste to save temples that sheltered no one but did nothing to protect their homes, which in their eyes seemed so much more precious.

The Nubians resigned themselves to their fate and, on leaving, asked to be pardoned by their ancestors, whose tombs were about to be covered by the waters. Seventeen temples were dismantled and reconstructed above the water line but many other identified sites were simply given a succinct description in files. Time was short and it became obvious that the research could not be taken further and that a large mass of information would be lost for ever.

Figure 80. The gold of the Pharaohs. Nubian delegation bringing gifts. From the tomb of Huy, vice-king of Nubia in about 1340 BC. At the end of the procession a nurse is leading two children and carrying a third in a basket on her back.

THE TWO TEMPLES OF ABU SIMBEL COULD HAVE BEEN SPARED

The project we have just described, however, leaves the two temples of Abu Simbel in the submerged area. Without going into the details of the Sudanese project, it is also legitimate to ask if, after the discovery of the Tushka Depression, it would have been possible to save from the outrage of being cut into segments with saws the giant statues of Ramesses II and his famous wife whom the river god liked to greet as he went past.

It would have been perfectly feasible to construct the dam a little further upstream of the two temples, founding it on a rocky outcrop on the right bank of a wadi called Wadi Or. The floodwaters would have been directed towards the large Tushka Depression by a canal of about fifteen kilometres in length cut into the left bank of the river (Fig. 81).

As the bed of the river is about 30 m higher than at Aswan, the dam could have been that much lower. In order to avoid damage to the temples by water seeping through the rather porous sandstone, each of them would have been first protected by an internal grout curtain, a technique in which the Russians had demonstrated their expertise with the enormous grout curtain under the high dam.

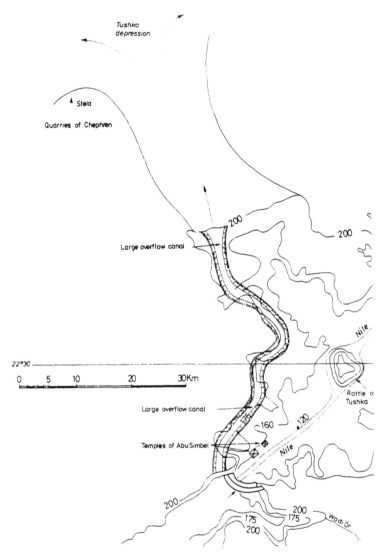

Figure 81. An even more rational project that would have saved the two temples of Abu Simbel.

Setting out from Aswan, cruise ships and feluccas would have sailed up the river through the charming valley of Lower Nubia. This journey towards Abu Simbel, the distant gateway to Egypt built by one of its greatest Pharaohs, would have had a powerful symbolic value. Then the boats could have ascended a shiplift at the side of the dam and made their way towards Sudan.

In the end the total cost would have been a lot less[13] and visitors of today would have been able to see the great king's masterpiece reflected in the water of the Nile instead of having first to look, from behind, at the hemispherical shells of reinforced concrete[14] covering the temples and then, on turning a corner, to find themselves suddenly face to face with the gigantic statues.

AN IDEA FOR ABU SIMBEL

In this connection I should like to offer a suggestion. One does not enter a theatre by the back door. All architects agree about one rule: to give a monument its full expression, it should be placed at the summit of a rising perspective. Forty-five centuries ago the great Pharaoh Cheops would have understood this perfectly: his pyramid, Akhuit the White, gleamed in the sun at the top of a monumental rising causeway nearly a kilometre in length. The effect must have been magnificent. Senenmut, the architect of the queen Hatshepsut, placed her temple at the very end of an avenue rising in a series of stepped terraces towards the entrance. Indeed, the temples were always designed with a gradual rise from the entrance at ground level to the holy of holies.

The great Ramesses, whose prestigious religious foundations along the valley in Nubia have been described by Christiane Desroches Noblecourt, has not really received the tribute he deserved. It would not cost much – compared with what was spent to rescue the temples – to set off the exceptional quality of this architecture at the top of a north-south rising perspective by building a new road with a series of levels or terraces bordered to the east by a belvedere overlooking the Nile. Two French architects, Philippe de Fouchier and Laurent Serraz, who share our views, have kindly made the drawings reproduced in Figures 82 and 83.

CONSEQUENCES OF THE HIGH DAM AS BUILT

The Egyptians bristle at criticism of their High Dam, which has become a national monument that reminds them of the fight to consolidate their independence. To realize this, one has only to glance at the documents in the archives showing the people's immense joy when the final tranche of the project was set in motion in the presence of Nasser and Khrushchev in 1964.

There was a lot of criticism abroad but nothing was said about what should have been done – a reproach that cannot be made to this study – and no account was taken of the problems that Egypt was trying to solve.

---

13. Not only would the dam have been cheaper to construct but there would have been no need to rescue the temples of Abu Simbel, which alone cost nearly fifty million dollars, or the other temples in the valley (nearly as much) or to displace the local population of 80,000 Nubians.
14. Recently masked by rockfill.

Figure 82. A project to enhance the setting of the temples of Abu Simbel: a new tree-lined road from the village to the west (above the temples in the sketch) turns north of the temples and arrives at the foot of a rising perspective towards the temples. (Ph. de Fouchier and L. Serraz, architects).

Figure 83. Arrival at the temples of Abu Simbel with a rising perspective. (Ph. de Fouchier and L. Serraz, architects).

Faced with a population that was increasing rapidly, indeed very rapidly at the time of the project, it was – and still is – absolutely vital for Egypt to expand the area under cultivation and to improve the productivity of its agriculture. Taking his inspiration from the achievements of the Pharaohs, Nasser could have imposed a development plan for the valley that strictly delimited housing zones and agricultural zones, strengthened the river's embankments and raised the level of the areas set aside for housing, as part of a general scheme that would require the building of a number of small dams between Aswan and the delta. At the beginning of Nasser's reign, this solution would have had logic on its side; but just after independence, such an agrarian revolution would have been very unpopular as it would have violated a number of entrenched rights. Nor would it have received assistance from abroad. In contrast, the High Dam became the emblem of the revolt against the big foreign powers that wished to impose their conditions.

The High Dam had three advantages:

1. It regulated the discharge, making it much easier to use the river for transportation and entirely calming the fears of those living in the cities;
2. It allowed the expansion of areas under irrigation, which were meant to increase by 20%. However, owing to the fact that part of the irrigible land has been swallowed up by urban sprawl, the actual increase is only 5%;
3. It increased the production of electricity. But here the disappointment was even greater: the country's installed capacity was to be multiplied by four,[15] but as the water level of Lake Nasser has never reached its maximum and has in fact varied considerably, the turbines do not function efficiently so that, today, only 10% of the 90 billion kW/h consumed by the country is supplied by the dam and, since the Nile no longer follows it former rhythm and has hollowed out its bed, part of that is used for pumping water or fertilizer production.

There are a number of disadvantages:

1. Firstly, as we have said, its vulnerability: the huge dam, with its massive reservoir, is like a powerful bomb threatening to destroy the country. If the dam were to collapse, an immense tidal wave would sweep down the narrow valley and cause an utterly unprecedented disaster;
2. A *euphoriant* effect: the High Dam and its reservoir are so gigantic that the Egyptians have come to believe that their country will have the right for ever, *ad vitam aeternam*, to all of the water carried by the Nile, as at the time of the Pharaohs. We shall soon see how dangerous such a belief is likely to be;
3. The annual loss of 10 billion m^3 of water by evaporation and infiltration;
4. The regression of the delta or, more precisely, the acceleration of its regression,

---

15. 1 900,000 kW instead of 500,000; see the IBRD report of 28 February 1956.

which had already started after the construction of the small dams below Aswan. What used to be dry land has become water again, as Ovid put it in his *Metamorphoses.*[16] This loss of territory is and will be increasingly sensitive in a country wedged between two deserts, whose cultivable land is only 3% of its total surface area;

5. The disappearance of the agricultural system used by the Pharaohs and growing salinity of the valley soils.

In former times, the soil was well aired and everything grew without effort after the inundation. The silt deposits have now been replaced by fertilizers, the excessive use of which has polluted and clogged the soil. Admittedly the silt was a poor fertilizer that contributed only about seven kilos of nitrogen per hectare but it also added a few salts to the water table – witness the sound made by the colossi of Memmon as they were warmed by the first rays of the sun.[17] The invasion of salts is a widespread phenomenon on fertilized soils: in the fields beside the Tigris and Euphrates, the hot and dry climate has caused the salts to rise to the surface by capillarity, with the result that the fields in Iraq contain so much that they sparkle in the sun. It was the same in the valley of the Indus. On the other hand, the salinization process was halted in Egypt by the rinsing and draining of the soil when the floodwaters subsided. The situation is quite different today following the construction of the dam, the harvesting of several crops a year, excessive fertilization and inadequate drainage.

The biological equilibrium has been dangerously degraded: organic matter, the crucial 'biological engine' that enables plants to take root and find nourishment, now forms only two per cent of the soil. When the amount of organic matter drops below two per cent the soil hardens and loses its contact with the air. If too much is demanded of the soil in the valley, it might cease to support the life of Egypt: whitish patches are already beginning to appear in Upper Egypt.

In the delta, an ecosystem is coming into being whose drawbacks are all blamed – unjustly in my view – on the High Dam. Algae and predators from the Red Sea have introduced a new underwater fauna and flora from Port Saïd to the delta: weeds and water hyacinth, which consume huge quantities of water, infest the canals and sardines and other fish that used to be common have disappeared, devoured by new predators. These developments are as much due to the opening of the Suez Canal as to the shortage of non-polluted fresh water in the delta. Too much water is consumed in the plain and the Nile now reaches the end of its long journey practically bloodless, like a giant pipeline tapped in too many places. It is polluted by the nitrates leached from the fields and by the waste water of the im-

16. *Vidi ego quod fuerat quondam solidissima tellus*
    Esse fretum, vidi factas ex aequore terras.
    I saw what was once the firmest of land
    Become liquid, I saw land made out of a sea.
17. A recent study has shown that this 'song' was caused by the fact that the sun dilated some of the salt crystals that rose from the water table by capillarity.

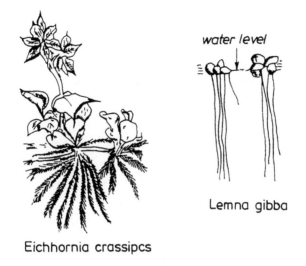

Lemna gibba

Eichhornia crassipcs

Figure 84. Water hyacinths, also called Nile roses, infest the canals.

mense megalopolis, while the fresh water that remains no longer has the strength to combat the salinization process. All this has less to do with the High Dam than with the considerable water needs for agriculture and the cities all along the valley. In the last hundred years, the population of Egypt has increased tenfold but the discharge of the Nile has remained the same.

Here we can see the consequences of man's vanity in seeking to tame nature: the current of fresh water reaching the delta used to prevent the predators from entering. As a supreme irony, large areas of water hyacinth, great consumers of water, infest the canals and absorb the meagre remains of the river, allowing predators to enter while in the stagnant pools bilharziosis is spreading.

Medium and long-term problems: silting of the reservoir

In the time of the Pharaohs, the civilization of Egypt owed its existence to a perfectly ordinary hydrological phenomenon: heavy rains tore shreds from the fairly fertile highlands which were then deposited far away in the valley. With his High Dam, Nasser has interrupted this process for ever.

In China, to return to our comparison, the damming of the relatively limpid Yangtse-kiang and of the turbid Yellow River are two very different problems in their long-term effects. The grains of silt torn by the Blue Nile from the Ethiopian mountains are relentlessly filling the reservoir with each season of floodwaters. No prior sedimentation test on a scale model was carried out and the sedimentation occurring at the upstream end of the reservoir came as a surprise. In the case of the Three Gorges Dam, on the other hand, the Chinese, after a series of tests,

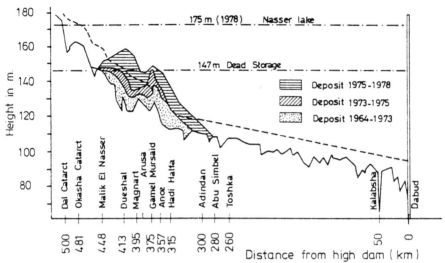

Figure 85. The silting of the reservoir in 1978, only fourteen years after the filling of the dam. Note how quickly the silt is advancing towards the dam (Egyptian document submitted to the I.C.O.L.D.), 1993, p. 311).

equipped the dam with twenty-three undersluice gates that made it possible to expel the turbid water after having collected the clear water. The High Dam, with its outlets too high and its excessive capacity that deadens the flow, acts as an enormous silt-trap. Figure 85 brings out the extent of the reservoir's dead storage, where the silt settles. In the medium term, silt deposition will work its way from the upstream end towards the dam; later on, the openings of the conduits leading to the turbines will become choked so that, when the available capacity of the reservoir has become too small, the dam will be destroyed by submersion. Let us be very optimistic and say that that moment will come in five hundred years time: in other words, during the five millennia separating Nasser from Narmer, it would have been necessary to reconstruct about ten of these enormous dams. Great rivers like to devour the barriers that are set in their way: in the bed of the Tigris, south of Mosul, the Iraqi scholar Nich has recently discovered the remains of three ancient dams built by the Assyrian kings in around 800 BC 'The waters of the Nile will never be halted', said the poet. The goddess Neith was right: when the dam collapses, the waters of the Mediterranean will withdraw from the delta, the delta will recover its shape and the silt will be spread over the valley as in the times of the Pharaohs.

The high dam is burdened with other reproaches that, in my view, are unjust. There has been talk of earthquakes induced by the weight of the water in the reservoir behind the dam. If that were to happen, it would have occurred shortly after the filling of the reservoir. Others have called attention to the creation of a micro-climate: there cannot be evaporation from the surface of a huge mass of water without causing a certain amount of humidity, but the consequences are not necessarily undesirable.

CHAPTER 5

The Nile, a major source of conflict in the twenty-first century: Egypt versus the upriver countries

'The waters of the river that rises in my country will be stopped from reaching yours, which I shall cause to die of thirst'.
The Negus of Ethiopia to the ruler of Egypt in the fourteenth century.

THE NEW NILE VALLEY, THE VALLEY OF HOPE: ANOTHER PHARAONIC DREAM

'What is done is done and cannot be undone'. So why cry over spilt milk? Will not a museum of Nubia suffice? My aim so far has been to underline the importance of reason and moderation. Now it is time to show how one pharaonic project has unfortunately led to another.

At first, in around 1959, the project to create a new valley involved simply the expansion of the oases of Kharga, Dakhla, Farafra and Baharïyya, whose subsoils contained layers of clay that trapped some isolated groundwater. The water obtained from about three thousand springs or wells from 30 to 80 m deep was supplemented by some three hundred boreholes to a depth ranging from 400 to 1500 m, but it was quickly discovered that the deepest boreholes tapped a water table of limited yield that was tepid and full of salts. Fourteen thousand hectares of cultivable land were irrigated by these means – a very small area in comparison with the three million hectares cultivated in the old valley. This result fell far short of expectations.

It was at this point, it seems, that the idea was born of creating a new continuous valley that would link Tushka to the string of oases further north. In other words, Tushka would serve not only as a receptacle for excess floodwater but also as the starting point for a project in which much hope was placed – a second Nile that would water a second valley, 'the parallel valley' as it was called. The Pre-Nile was rising from its ashes. But the whole scheme was based on a confusion between the *unpredictable* nature of the inundation and the *permanent* need to feed the population.

THE DIFFICULTIES OF THE NEW VALLEY PROJECT

When the Tushka overflow system first came into operation in 1987 on the left bank of Lake Nasser, the water flowed for a couple of weeks, carrying the fish with it and delighting the waterfowl before disappearing into the burning sands of the desert. It was then that someone had the idea of a pumping station that would feed in all seasons a canal linking the string of oases along the long-vanished Prenile: the enormous reservoir of the high dam had convinced the Egyptians that the water of the Nile was inexhaustible. So, why not use this water to create a new valley?

A first section of 350 km, initially called the Sadat and then the Zayed Canal, after a generous Maecenas of the time, Sheikh Zayed Bin Sultan al-Nahayan, President of the United Arab Emirates, would go as far as the Kharga oasis. From there, it would be extended to Dakhla and Farafra. Carried away by their enthusiasm, the planners continued their lines on the map towards the enormous Qattara Depression close to the Mediterranean, justifying the even more ambitious label – the 'new delta' – given to the project (Fig. 86).

The need for more cultivable land was so evident that the project was given an enthusiastic reception. When Egypt's cultivated land area is divided by its population it emerges that the surface per person has fallen from 70 ares[1] in antiquity to 50 in the nineteenth century and to no more than 5 today. The situation was explosive, since the annual increase in population was still over one million persons a few years ago.

'Let the desert blossom', said President Mubarak in 1997 as he launched at Tushka, thirty-six years after the laying of the first stone for the high dam. The Egyptian press have nicknamed it 'the inverted pyramid of the year 2000' – yet another reminder of the Pharaohs and their grand schemes. 'New cities will be created and new power stations built', added the Prime Minister.

Criticism did not take long to emerge. It was deplored that the preliminary studies, unlike those relating to the Peace Canal (about which more later), had been overhasty. The highly respected geologist Rushie Saïd, today Professor at an American university, was particularly severe: *'The authorities are conducting a public relations operation. Agriculture under the Tropic of Cancer is a waste of energy – the climate is too dry, evaporation too strong and the soil too permeable'*. In other words, he thought it highly improbable that the desertification process could be reversed. The site is close to the desert, where the winds, as violent and cold in winter as they are hot in summer, displace the sand dunes, which then block the flow of the water and lay bare the roots of plants.[2] Some people recalled

---

1. Butzer estimates that, under the Old Kingdom, a population of 1,100,000 persons cultivated 800,000 hectares.
2. B. Bousquet, 'Tell-Douch et sa région', IFAO 1996.

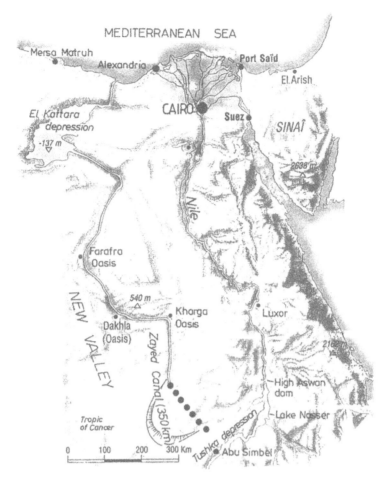

Figure 86. For the Zayed Canal it was needed to pierce a tunnel (shown in large dots) under the limestone plateau.

that the oases had been used a number of times in the course of history as a place to banish criminals and other undesirables.

Under the Tropic of Cancer, with summer temperatures exceeding 50° Celsius, nothing is easy. One particular problem will be to prevent leakage by laying a screen of polyethylene on the floor of the cutting – which will be difficult to make flat – and then to cover it with a thin layer of concrete to avoid damage to the plastic. Other criticisms followed: how would the canal cross the higher ground? How would evaporation be avoided? Would they adapt the old Persian method of ganats (undergound irrigation channels) and take the water under the hills? When one of the persons responsible for the project was questioned about it, he recognized all these problems and stated that the amount of water needed would make it necessary to install seventy-six underground pipes some two hundred kilometres long and that the cost of the operation would be three times the anticipated cost of the partly underground 350 km Zayed Canal project as far as Kharga.

The Egyptian intelligentsia did not cast doubt on the idea of the project but took a rather abstract approach, speculating about the number of feddan[3] that would be won from the desert, on where the project would end and on ways of coping with a hostile nature. The price of a barrel of oil having dropped, the generosity of the emirs faded away; today there is talk of a 20% contribution by the state with the rest of the cost to be borne by the new colonists. But as there are few takers, plots of land are granted free of charge to volunteers whose mental determination and physical resistance are tested by the authorities.

The idea is still alive. It is said that 'Many are the dreams that have become reality through diligence and perseverance',[4] and people think they should put their trust in the progress of science and technology. When you ask the average Egyptian what achievements his country can be proud of, he will mention the second valley and the Peace Canal. There is a bit of Gerard de Nerval in the Egyptian

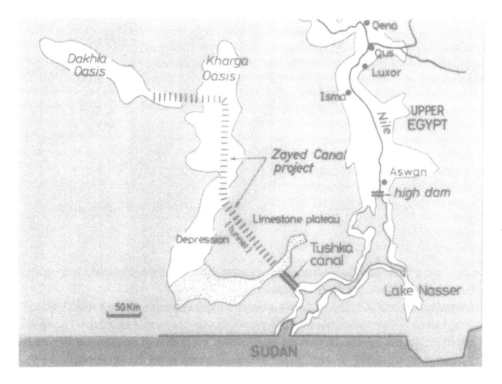

Figure 87. The Zayed Canal project with its underground section, is already abanaoneu in ravour of another project for an open canal that would wind round the western edge of the limestone plateau.

---

3. A feddan is just over an acre.
4. Ibrahim Nafie in Al-Ahram, 17 February 1998.

mind: 'They will come back, the gods you still lament! / Time will restore the order of former days; / The earth has shuddered with the breath of prophecy'.[5] The civilization of the Nile plain is not totally fatalistic – there is a deep-rooted belief that the world can be saved by creative intervention.

The Egyptian press, however, has not been entirely taken in concerning the problems encountered. The construction of the Sheikh Zayed Canal was recently at the centre of a polemic started by the daily newspaper *Al-Wafd*, which castigated the serious technical problems. Today, it seems that the idea of an underground section has been abandoned in favour of taking the canal round the western rim of the rocky plateau (Fig. 87), which would lengthen the route by about two hundred kilometres; but the new route passes through some very low-lying areas that would have to be filled in at great expense. For the moment, the project is limited to the irrigation of a few areas near Tushka, around lakes that are reminders of the Nile's diverted floodwaters.

THE EL-SALAM OR PEACE CANAL, A CURIOUS OFFSHOOT OF THE NILE'S HYDRAULIC SYSTEM

Rivers flow from their sources to the sea, receiving the tribute of their affluents, themselves swelled by innumerable streams, on the way. Today, in the case of the Nile, this law has been flouted. Not a single affluent swells its waters during the entire length of the plain it traverses on its way to the Mediterranean, yet it has been decided to stretch the river in two directions – towards the new valley mentioned above and eastwards by means of a canal towards the Sinai plain (Fig. 88).

The idea for this canal arose after the signature of the peace treaty between Israel and Egypt in 1979. Sadat wanted to bring the water as far as Israel. The canal was, as it were, an olive branch.

It was a bold technical challenge: to turn the yellow sand of the peninsula into fertile land and transform a stretch of desert into a green and prosperous region with several million inhabitants.

In 1991, the canal was financed by a loan of one billion dollars from the Kuwait Economic Development Fund and called the Sheikh Jaber Canal, after the Amir of Kuwait. But the plan to take it as far as Israel was soon dropped and the terminus of the canal became El Arish on the Mediterranean coast. However, even to reach that port, it was noticed that the hills in the region would make it necessary to build very costly pumping stations. As things stand, the Sinai is still waiting for its water and only 86 km of the project have been completed, as against the original plan for 242 km.

---

5. From 'Delfica' in *Les Chimères*: 'Ils reviendront les dieux que tu pleures toujours! / Le temps va ramener l'ordre des anciens jours; / la terre a tressailli d'un souffle prophétique'.

Figure 88. The Peace Canal, final branch of the Nile's strange hydraulic system.

The Pharaohs, as we have seen, wanted to surpass their predecessors but at the same time show their respect for the wisdom of the past, and in particular its deep understanding of water. One did not start the construction of a canal without having first carefully determined its 'horizon' – and the degree of precision implied by this notion is illustrated by the care taken in levelling the rock at the base of the Great Pyramid which, over a distance of 220 m is perfectly horizontal to within just over a centimetre. In the modern era, the planners of the new valley and the Peace Canal do not seem to have attempted such a high degree of precision!

The Egyptians have given themselves up until 2005 to terminate the work, but enthusiasm is waning because irrigated cultivation is very hard work. North Sinai raises problems since the land is often gorged with salts and will need prolonged rinsing. The government is holding back the water from the canal so long as investors use it for fish farming, which is more profitable; the authorities rightly consider that this costly canal and its pumping stations had not been built in order to breed fish. They have decided to combat this nonchalance – and the speculation in land – by inserting into the contracts a clause under which the buyer, if he continues to baulk at cultivating the land, forfeits his rights of ownership.

EGYPT'S WATER NEEDS

CHANGES IN THE AGRICULTURAL ECONOMY

The Gross Industrial Product of Egypt has quadrupled in the last few decades but has risen less quickly than the population. Egypt is still a predominantly agricultural country and, although progress has been made, the nation is far from being able to feed itself. At the beginning of the eighties there was a saying – probably somewhat exaggerated – that fourteen out of fifteen loaves of bread sold in the country came from abroad. At the time the massive importation of flour was regarded as one of the most important problems of the economy. Since then wheat production has increased, but more slowly than the population, and 45% of needs still has to be imported.

Since Mehemet Ali there has been an expansion of cash crops. The trend started with cotton, now cultivated on a fifth of the country's usable land. This cotton, of excellent quality, was for a long time the only true export product. It was followed by rice, maize and sugar cane, with the result that less attention was paid to subsistence crops, leading to the deficit mentioned.

At the start of the twenty-first century the vital question is whether the country's water resources will or will not enable it to satisfy all its ambitions for the production of food and cash crops. There is not the slightest chance of the country becoming once again, as it used to be long ago, the granary of the world, since it is estimated that the population to be fed, sixty-three million today, will reach a hundred million by 2030.

THE RULE OF THREE, KEY TO EGYPT'S AGRICULTURE

In the nineteenth century, Linant de Bellefonds, during his long life in Egypt, did not fail to compare the discharge of the Nile and the area cultivated. He deduced from this the amount of water needed for agriculture: according to his calculations, the river needed to spread 1.1 m^3 of water over each square metre of irrigated land in the course of a year. In 1993,[6] for a cultivated surface of 3.5 million hectares, water consumption came to 51.6 billion m^3, raising Linant's figure to 1.34. The rise was due to the practice of perennial irrigation of rice and sugar cane, both of which use large amounts of water. This relative stability of the criterion put forward by Linant shows clearly that the Nile can feed on its way through

---

6. M.N. Ezzat, 'Nile water flow, demand and water development', Minutes of I.C.O.L.D. 1993.

Egypt only a limited number of human beings, since each individual consumes plants and cereals that require for their growth a great amount of water.[7] Egypt has already given up all hope of self-sufficiency in food; it could even be said that the country is a big importer of water – the water needed to grow the food it imports in the country of origin.

TOTAL ANNUAL NEEDS

In addition to the 51.6 billion m^3 used for agriculture, the annual domestic demand in the cities, towns and villages is 4.5 billion m^3 of water, the needs of industry 4.7 and of navigation 1.8; all in all, taking into consideration the extension of cultivated areas in recent years, the country now consumes some 66 billion m^3 of water.

THE 1959 WATER-SHARING AGREEMENT BETWEEN EGYPT AND SUDAN

Under the terms of this agreement, concluded just before the construction of the high dam, the water shares attributed to the two countries by themselves were respectively 55.5 and 18.5 billion m^3 out of a total flow estimated at 84 billion m^3, the other ten billion corresponding to losses from evaporation and infiltration. In other words, Egypt already consumes 10 billion m^3 more than its agreed quota. This situation is possible only because Sudan does not yet consume its own full share and because the discharge of the Nile was slightly underestimated and losses overestimated.

IMPACT OF THE NEW VALLEY PROJECT

According to a recently published document,[8] the purpose of the new valley project is, in the space of twenty-five years, to multiply by five the present area under cultivation. The wishful thinking in this project is immediately apparent: agriculture alone would consume 257.5 billion m^3 (51.5 × 5), that is more than three times the total discharge of the Nile. Other more realistic estimates limit the ambitions of the project to a 25% increase, but even that would require an extra 12.8 billion m^3 of water, which would bring the annual consumption of Egypt to 74 billion m^3 and leaves nothing at all in theory for Sudan. Water nourishes pipedreams and figures are bandied about with little attention to their impact on the treaty signed.

The water issue is of crucial importance for Egypt and will be difficult to resolve, for three reasons:

---

7. So much so that, in the end, a tonne of dry wheat requires 1000 tonnes of water.
8. 'Egypte-France, les Ponts de l'amitié', a document distributed by the press service of the Egyptian Embassy in Paris.

1. The possible ways of saving water in the old valley are very limited:
   One could, of course, improve certain bad habits in the cities and repair leaking pipes, but that would concern a sector that accounts for only a few percentage points of total consumption. Indeed, certain underequipped urban areas have no piped water at all and the fetching of water is a painful chore for many women. (In Africa, it is estimated that a woman spends on average six hours a day obtaining the precious liquid.)[9] The main potential savings are in agriculture and it is rightly envisaged to outlaw irrigation by submersion and to replace it by drip irrigation (already used in Israel, Jordan and the United Arab Emirates), but the changeover will not be completed until the second decade of the new millennium and then only for a third of the area under cultivation.Another idea is to halve the area under rice, which requires the most water, but this would have the drawback of increasing imports that are already high and of depriving the peasants of their most profitable crop.

2. The current projects aimed at increasing the flow of the Nile will remain purely theoretical for a long time:
   The biggest one is the digging, in the Sudd in Sudan, of a canal called the Jonglei, some 360 km long, which would shorten by 300 km the length of the White Nile and avoid all the evaporation that occurs as it makes its way slowly through the region (Fig. 89). The savings in water are estimated at a little over 4 billion m^3 a year, equivalent to less than half the losses through evaporation on Lake Nasser.

   Work was started on the project in 1978 and continued until 1983, but has now been interrupted indefinitely by the civil war between the north and the south of Sudan. John Garang, the leader of the rebellion of the South, condemns the project for both technical and political reasons. The future of the project is clearly uncertain. Not only will it be difficult to carry out but it will in no way satisfy the vital needs of tomorrow's Egypt.

   Other projects are under consideration, such as increasing the capacity of the equatorial lakes, excavating even longer canals than the Jonglei and raising embankments along each side of the White Nile to give a greater flow in all seasons. All are gargantuan schemes that will require enormous investment and full agreement between the countries concerned. If it is really desired to carry out these pharaonic projects, and to do so without ecological damage, they will need to be studied very thoroughly: there is a price to pay for diverting a river from its course. The meanders of the king of African rivers, like those of all rivers, obey their own laws.

   Another scheme is to recycle the waste water of Cairo, but that would have only a small impact; pumping fossil water from deep reservoir rock is no more promising. In the end, the only solution near the coast is the desalinization of

---

9. France Bequette, *UNESCO Courier*, June 1998.

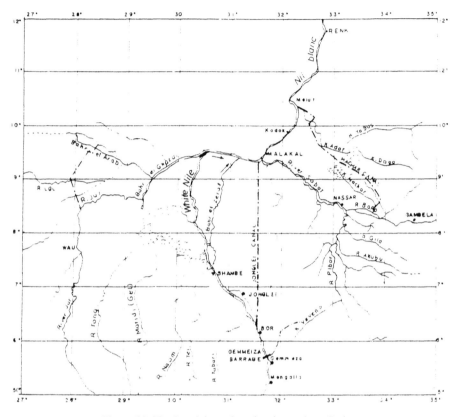

Figure 89. The Jonglei canal project in southern Sudan.

seawater which, despite recent technical progress, still costs about a dollar per cubic metre.

3. The water needs of all the countries bordering the Nile are increasing: their total population amounts to three hundred million persons and, if United Nations figures are to be believed, will reach one billion in about fifty years. Set against the demographic landscape of tomorrow, the situation of Egypt and the upriver countries is a matter of grave concern.

THE TUSHKA PUMPING STATION AND THE
ANGER OF ETHIOPIA

For a treaty to remain valid, things must stay as they were – 'rebus sic stantibus' as the Romans put it. Ethiopia and the other upriver countries already refused to recognize the Egypt-Sudan treaty of 1959 and did not fail to denounce with the utmost vigour the recent ambitions of Egypt regarding its new valley and Peace Canal.

For a moment they had hoped that the new valley project was for the most part wishful thinking. But in February 1988, they found out indirectly that letters of intent had been sent by Egypt to the United Kingdom-based Kvaerner International and Japan's Hitachi Corporation concerning the delivery, for a sum of 425 million dollars, of a pumping station with thirty-one pumps capable of extracting 5.5 billion m^3 of water a year from Lake Nasser just opposite the Tushka Depression. There could no longer be any doubt: the project was not just an overflow system for flood waters. And so, in March 1988, at an assembly of states bordering the Nile held in Tanzania, Ethiopia and Uganda emphasized the iniquity of the 1959 treaty and asked for its abrogation and the negotiation of a fresh treaty. They fixed in advance their own shares at 18 billion m^3 a year each.

The matter has been laid before the Organization of African Unity. In response to the request for the abrogation of the treaty presented by the Ethiopian Minister for Foreign Affairs, Egypt signalled its agreement to discuss the issue in order to clarify this '*misunderstanding*'.

THREATS FROM SUDAN AND ETHIOPIA: LEVIES AT THE SOURCE

The Pharaoh is the king of the waters: it is he who gives water to the land, we are told by the texts. Why should this right be refused to the upriver populations?

Why should Sudan, threatened with creeping desertification from the west, not make every effort to expand its irrigation in the rich plain of Gezira, the 'Land between two Niles',[10] and in the rich province of the Blue Nile to the east, where the mechanical cultivation of sorghum and sesame is being developed? Evidence of this irrigation policy can be found in the construction of the Roseires Dam for which Sudan got the go-ahead in the 1959 agreement, in compensation for the building of the high dam (Fig. 90). Do people really believe that, if the Jonglei Canal ever becomes a reality, Sudan will refrain from using at least part of the extra water thus obtained to irrigate regions in the surrounding area? Further south, the Sudanese are planning to build a number of small dams (Fig. 91) for irrigation schemes that will reduce the flow of the White Nile. Even further upstream, Rwanda and Burundi are considering a dam on the Kagera.

For many people, dams are harmless things that would have no impact on the overall discharge of the Nile. It is perfectly true that a dam for the production of electricity does not diminish the flow of a river: it simply regulates that flow. But the dams in question here are for irrigation and each one of them is a 'bloodsucker' in the Nile's tropical reaches pumping the liquid nourishment that used to flow down towards the plain. An upriver population will then develop and make the process irreversible. There will no longer be any talk of what Julius Caesar

---

10. P. Crabbe, 'Alluvium plains', in *A Land between Two Niles*, Balkema 1982, p.5.

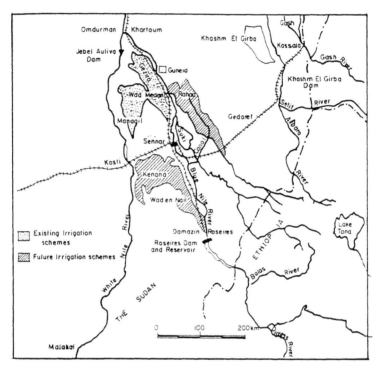

Figure 90. Irrigation schemes in north Sudan, from an Egyptian document submitted to the
I.C.O.L.D., 1993.

Figure 91. Some existing and projected dams on the Nile in Sudan, Ethiopia and Uganda.

called 'the long series of yearly inundations of the Nile' but rather of a breast exhausted by the suckling of too many children.

On the face of it, the Ethiopian threat is even more serious. In May 1998, an Ethiopian governmental committee announced its intention to build a large number of dams on the Blue Nile in a study that, according to the Minister for Foreign Affairs, covered 44 dam sites and 132 irrigation schemes. Ethiopia was seeking to transform its agriculture by changing over from a system based on wells to one based on irrigation by immersion with the purpose of developing millions of hectares of land – a lot to ask in such a mountainous country. It also wanted to profit from the nature of its terrain to produce electricity.

The President of Ethiopia, wishing to explore in greater detail the possibilities open to his country, has commissioned studies from firms independent of the World Bank – the French firms B.C.E.O.M.-Eau (Ch.Pontier) and I.S.L. for the Blue Nile basin and two Anglo-American firms for the basins of the Omoi-Gibe (an affluent of Lake Turkana) and the Sobat (an affluent of the White Nile).

The greatest potential resources for energy production are those of the Blue Nile, which drops about 1300 m between Lake Tana and the frontier and has a catchment area half the size of France. Along the course of the river there exist four potential dam sites with drops, when the reservoirs are full, that would produce 5000 megawatts, the equivalent of five medium-sized nuclear reactors. The country as a whole could produce double that capacity, though its own needs are not expected to exceed 1000-1500 megawatts in the future.

This offers a rich potential for export, and the dam reservoirs would totally regulate the discharge of the Blue Nile.[11] But the construction of these dams will require considerable capital investment. The first of the four, the Karadobi Dam for example, would be no less than 250 m high. But the scheme would regulate the Blue Nile and provide northeast Africa with a source of renewable energy.

On the other hand, the systematic development of irrigation on the high plateau will have to overcome a number of problems: entrenched traditions, an unhelpful soil and an uneven terrain, even on the high plateaus, that will make the irrigation dams expensive to construct. In short, a number of projects are maturing that are likely to modify the flow of the Nile before it gets to Egypt. Africa is an arid continent where water is becoming increasingly scarce; according to the recent meeting of experts in Rome in connection with the Convention on the fight against desertification, it is the continent most affected by that process. The Nile is neither 'swelled by the rains of Zeus',[12] as Homer wrote, nor born in paradise, as the chroniclers of Saint Louis' crusade would have it. There will be no miracle: the Nile is not an inexhaustible manna from heaven.

One could go even further. Some eight or nine thousand years ago monsoon rains watered the South Arabia peninsula and what is now the desert of Yemen

---

11. The total capacity of the reservoirs of the four dams is estimated at 64 billion m³.
12. The Odyssey (Book IV, line 581).

was then an area of green that used to be called 'Arabia Felix'. Since that time, however, the moisture-laden winds have shifted their rendez-vous to Ethiopia. But will it always be so? We have already noted how fertility-bearing monsoons once, long ago, greened certain parts of the desert at Nabta, west of Tushka. Will the present monsoons continue to pour down from the Ethiopian skies or will they move somewhere else? Will we speak in some very distant future of an 'Aegyptus Felix'? We know so little about long-term trends in climate.

When all the projected irrigation dams have been built along the upper reaches of the river, they will be the first to fill up when the river swells, leaving Lake Nasser to take what is left over. What then will remain of the enormous reservoir, already burdened with silt, from which water will have been pumped all through the year to feed the new valley? One is reminded of the tragedy of the Aral Sea in Siberia, which used to be fed by generous rivers,[13] along which, however, the Russians have developed the irrigated cultivation of cotton. As a result the Aral Sea is dying.[14] Lake Nasser will become another Aral Sea, with its waters gradually dropping to uncover vast quantities of silt.

THE PROBLEMATIC RENEGOTIATION OF NILE WATER RIGHTS

The bipartite agreement of 1959 was not the first of its kind. There had been one signed by Egypt and Great Britain in 1929 under which Egypt had been attributed 40 billion m³ and Sudan only 4. In 1959, the Sudanese, as we have seen, had made it perfectly clear that they repudiated an agreement they had not signed and ended up, under the new treaty, with a much bigger share.

Ethiopia then expressed its astonishment that Egypt and Sudan could have deemed it possible to share between them a 'cake' three quarters of which was supplied by its territory and regarded – and continues to regard – that treaty as null and void. 'The present situation is not only unjust but totally unacceptable', insisted Seyum Mesfin, the Ethiopian Minister for Foreign Affairs.

The United Nations Economic Commission for Africa sounded the alarm in 1995: 'If no action is taken to increase the flow of the Nile and employ more efficient irrigation methods, there will be a disorderly rush by riparian states to grab the water of the Nile for themselves. The situation could degenerate to such an extent as to provoke open conflict between these countries'.

It would seem at first sight that the protests and warnings have not been totally in vain, since the responsible Ministers of all the countries concerned agreed in May 1999 to exchange views in Addis Abeba.

---

13. They used to deliver some 50 billion m³ of water a year but now deliver only 10, the other 40 being absorbed by massive irrigation schemes.
14. It is now half the size it was and what remains is polluted by heavy metals, fertilizers and insecticides, and other toxic substances that are seriously endangering the health of the local inhabitants.

THE NEED FOR A NEW MENTAL OUTLOOK

These self-assertive protests from countries most of which had been vassals of ancient Egypt shocked the feelings of the Egyptians: in former times the Pharaoh, master of the living and the non-living, had the right 'to use and abuse' water. A conservative outlook: what the river gave should continue to exist. Times appear to have changed. Upriver there has been a revolution in thinking: some of their ancestors had been slaves, but the descendants of those slaves wish to be the masters of the water that is born in their country and which they see flow past them every day. They know perfectly well that the water belongs to nobody, that it pays no attention to frontiers or to religions, and that the owners of the upper river basin are always in the strongest position. They protest against the egotism of the downriver countries, who are determined to hold on to their privileges.

The difficulties of reaching agreement are political, ethnic, religious and cultural in nature. The territorial limits are far from being what they were and certain upriver countries conceal irredentist or even imperialist ambitions.

Uganda dreams of becoming the pivot of a regional entity grouping together the pastoral communities of eastern and central Africa.

Sudan, the largest country in Africa, is an artificial unit stretching from south to north under a severe climate. Its total population of only 31.5 million inhabitants contains about twenty different ethnic groups and nearly 600 sub-groups, with Arabs making up 40% of the whole. Right from the proclamation of independence in 1956 the South, peopled by black-skinned Christian and animist tribes, rebelled against the North, whose population is 75% Arab-speaking Muslims. The 1972 agreements accorded self-government to the South on condition that they renounced their separatist aims. In 1983, however, the proclamation of the Sharia, that is to say the application of Muslim law, throughout the country rekindled the war, which is still going on, ruining the country and greatly reducing hydraulic investments in both the north and the south. The army of the Sudan People's Liberation Movement (SPLM) and its leader John Garang have not the least intention of facilitating the supply of water to the north and they have brought to a halt the construction of the Jonglei Canal which Garang claims, apparently not without reason, will drain the Sudd and turn the whole area into a desert, kill off the livestock and destroy the ecological equilibrium. He made the whole issue the subject of a thesis which he submitted to the American University of Iowa.

North Sudan is tempted to foment murky plots by fundamentalists and terrorists, heirs of the celebrated 'Mahdi' Mohammed Ahmed, who defeated the British in 1885 and killed General Gordon at Khartum. The Khartum junta is faced with a catastrophic financial situation. However, a new opportunity for the country as a whole to get richer has arisen: large reserves of oil have been discovered in the southwest, at Bentu, where a dissident faction of the SPLM holds sway. Seven thousand workers and engineers have just completed the construction of a 1600 km pipeline linking Bentu to Khartum and Port Sudan and the very first exports

Figure 92. Bentu and the Sudanese pipeline under construction.

of crude oil left for Singapore at the end of August 1999 in a first step towards industrialization. Marketing is in the hands of an international consortium in which China and Malaysia have a majority holding. It is a riposte to the hostility of the USA. For a long period Egypt took delight in stirring up trouble between the North and the South. The stubborn resistance of the SPLM is leading it to reconsider the situation: an independent Christian and animist south Sudan could fall under the influence of the USA, Israel or other countries in the region and endanger the waters of the White Nile.

Ethiopia has the biggest population (57 million inhabitants) and is poor but with a slowly growing economy. The country is still remembered for the myth of King Solomon and the Queen of Sheba. Nearly a thousand years after their legendary meeting, which produced the *negusa nagast* ('king of kings'), Menelik I, founder of the Ethiopian Church and the person who transferred the Ark of the Covenant from Jerusalem to Aksum, Strabo wrote: 'The Ethiopians live a nomadic life and have few resources because of the sterility of their land... while the Egyptians are in quite the opposite situation' (Book 3). This remark about the sterility of the highland country, probably quite accurate at the time, is deeply entrenched in the Egyptian mind: for them, the Ethiopians have no needs. The Egyptians haughtily disregard the Chinese proverb: *when you drink water mind you of the spring*. And yet in the past, all the peoples of the Nile regarded the Abyssinians as the 'masters of the waters', those who had the power to check its course or to make it flow, not from east to west, but in the opposite direction; for the Muslim Pasha of Cairo, the very Christian Negus was an enemy who had to be treated with respect on account of his power over the Nile. But the Egyptians eventually forgot about all that. What has recently angered the Ethiopians is the wastage of the waters of their Nile by the Egyptians, who are pouring it into the sands of the Tropic of Cancer. It called to mind an old Abyssinian proverb: 'When people loot the house of your father, join the looters'. The house of the father is the Nile; it is being looted downriver; in their uplands, the Ethiopians, too, are going to make much greater – and even unreasonable – demands on the river since the Blue Nile, the Atbara and the Sobat have relatively short courses in their mountainous country.

For the Ethiopian Orthodox Christians, Lalibela, a site high up in the mountains with a series of churches hewn out of the rock where diluvian rains fall from June to September, is a second Holy Land, a second Jerusalem, with a Mount Tabor at the foot of which runs a stream, renamed the Jordan, that is the beginning of the Atbara. This water will descend towards two predominantly Muslim countries. The Emperor Menelik II (1865-1913), renovator of the country and victor over the Italians at Assoua. said 'Ethiopia is an island in an ocean of infidels'.

Lower down, Muslim Sudanese Nubia was once the powerful Kingdom of Kush, rich on account of its gold, where two hundred pyramids, twice as many as in Egypt, are to be found. Its people remember the humiliations suffered and the tributes in gold and human beings exacted by the Pharaohs. In 836 AD, Nubia was forced under a treaty to provide each year *three hundred and sixty slaves of both sexes, the finest-looking in the country, with no decrepit old man or woman and no boy or girl under the age of puberty.* Long ago, the region which controlled the lines of communication was bled of its best elements. Why should it refrain today from controlling the water of the river?

Religion lends its weight to power and each country has its own views about the development of its population and the use of its natural resources. The Muslim religion, for example, in reciting the blessings of Allah, asks 'Is then he who creates like one that creates not?' Such an injunction probably has something to do with the building of the high dam and the creation of the new valley. And in the rising population of Egypt, so worrying in relation to water resources, is there not some inspiration from the sixth Sura of the Koran on 'Cattle': 'Kill not your children for fear of want; we shall provide sustenance for them as well as for you', and the Koran expresses well the future problem of man and the Nile: 'Allah created all life with water but he set against it the stones and the sand of the desert so that the elements, like human beings, will use their strength against one another'.

Under the Pharaohs, the peasants of the plain had learned that water was a matter of sharing. The cuttings in the embankments of the Nile ordered by the Pharaohs were the expression of a grand design for the community: the fellah received the water due to him and took no more because he knew that to do so would harm his neighbour. What would have happened in a divided Egypt, if Lower Egypt had had ambitions of this kind? Upper Egypt would have responded by wasting the water and the silt and the North would have to make do with what was left over.

The problem today is to give an international dimension to this long-established ethic of water. Certain rivers, such as the Mekong, the Paraguay and, in Africa, the Senegal, provide examples of successful transfrontier agreements but, elsewhere, how many sources of tension are due to problems of water! The problem is not new but it is getting worse and along the Nile it is particularly acute because of the particular history and geography of the plain of Egypt: almost no other country on the lower course of a river is as dependent on the river as Egypt is. And this land of a most ancient civilization, whose very survival is at stake, has a place in every person's heart.

In order to persuade the Israelis to give back the West Bank and Gaza, Sadat offered them the water of the Peace Canal, and in doing so aroused indignant protests both in Egypt and in the upriver countries: the latter justifiably argued that Sadat had no right to give away the waters of the river without their agreement. But this did not stop Sadat, a little later, when Mengistu, 'the red emperor of Ethiopia', signalled his intention of using the water of Lake Tana for irrigation in his country, from threatening military action: 'Our life depends 100% on the Nile and if anyone, at any moment, has the idea of endangering our lives, we shall not hesitate to go to war, for it is a question of life or death'.[15]

As often happens in similar circumstances, people will refuse to recognize the urgency of reaching an agreement in the hope that Africa will always be Africa – *Africa semper Africa* – and that the many questions dividing the upriver countries will not fail to weaken their demands. Egypt will do nothing to attenuate these divisions and, faithful to the old maxim of keeping on good terms with the neighbour of one's neighbour or of making a friend of the enemy of one's enemy, recently extended its welcome in Cairo, in December 1997, to John Garang. It also supports Eritrea and Somalia in their conflicts with Ethiopia. At the same time, Khartum has recently declared itself to be the best friend of Kabila, who is himself fighting Uganda and Rwanda. But all these manoeuvres more or less cancel each other out while, behind the scenes, Israel acts with much greater realism.

ISRAEL, THE ARAB LEAGUE AND THE ZIONIST LOBBY IN THE UNITED STATES

Israel has forgotten the olive branch offered by Sadat; it remembers only his threat to Ethiopia and his assertion that Egypt depended 100% on the waters of the Nile and would not hesitate to take military action. Ethiopia has always been a focal point in the regional strategy of the Hebrew state, which is well aware that it can use water to weaken Egypt and thus neutralize the 'Arabo-Islamic' encircle-

---

15. B. Boutros-Ghali, *Egypt's Road to Jerusalem.*

ment. Ethiopia had no choice but to subscribe to these views and in 1957, just after the Suez crisis, it authorized the opening of an Israeli consulate in Addis
Ababa. Diplomatic relations were resumed with Mengistu in 1989.

Israel then began to act on many fronts. To start with, it sent agronomists and
irrigation experts to Ethiopia and beyond, to South Sudan, Uganda and Rwanda,
who introduced techniques in which their country was very experienced; in particular, they chose the sites for many dams. Since the Camp David Agreements,
of course, the presence of these experts has been presented as part of the quest for
economic development. In Ethiopia, the need to call in foreign experts is greater
than elsewhere because many large properties are fragmented by deep valleys that
cut them off from the rest of the country. The aim is to regenerate the forests and
amalgamate the large number of microsocieties over which the emperor – the
'atse' – used to reign.

But Israel has also provided military support to these countries, including the
provision of weapons, military instructors and training courses in Israeli army
colleges.

The third aspect of their action, though related to the other two, is more intellectual in nature: they are teaching those countries how to levy water at the
source, taking their cue from recent developments near the sources of two other
rivers in Turkey. On the other side of the Red Sea there exists a great river, the
Euphrates – 'the river that flows back to front' as the Pharaohs used to say. It
originates in Turkey, like the Blue Nile in Ethiopia. In 1992, without making the
least agreement with Syria and Iraq, the Turkish Prime Minister asserted the total
sovereignty of his country over the waters of the Euphrates and the Tigris, regardless of the consequences this might have on Turkey's downriver neighbours.
He then decided on the construction of the Atatürk dam, which contains 50 billion
m^3 of water. This irrigation scheme is transforming the south of Turkey, which
has become a garden of nearly a million hectares (Fig. 93).

The Israeli experts are well aware of the great fertility of the Ethiopian volcanic
soils when they are watered and they have taught the local peasants the art of dis-
tributing this water in a large number of upland areas before it enters the deep
valleys of the Blue Nile and its affluents. Then will the blossoming of the Ethiopian orchards foreshadow the wilting of those which the Pharaohs in the plain
used to call the 'orchards of Osiris'.

One day soon Ethiopia, like Turkey, will firmly defend the absolute principle
of its sovereignty over the upper catchments of all its rivers, while Egypt will
continue to argue for the imprescriptibility of its acquired rights. The situation
will rapidly become explosive as it is at the centre of a conflict between Israel and
the Arabs. The policy of Israel appears to justify the cruel aphorism of Professor
Raymond Weill: 'The water is cut first, then throats'.

The recent visit to Ethiopia and the region of the Great Lakes by the then
American Secretary of State Madeleine Albright also confirms a zionist infiltration in these regions with, under the cover of democracy and cooperation, the

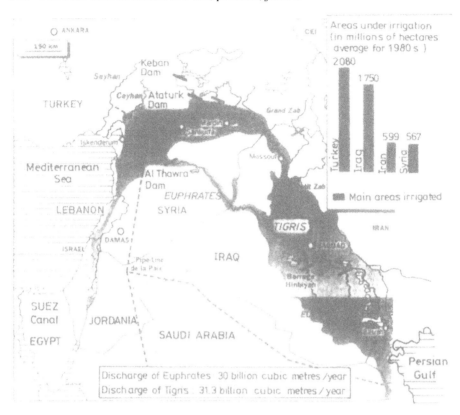

Figure 93. Areas irrigated by the Euphrates and the Tigris. At the top, the areas of Turkey irrigated since 1992 by the Atatürk Dam, which reduces by 50 billion m^3 a year the flow of the river in Iran and Iraq (World Media Network).

tions for water and the reduction of Egyptian quotas. No more innocent is the submission to the United Nations by the United States Congress of a draft international convention under which all nations would be free to construct dams on their territory for the purpose of promoting economic development. It is a text that seems to have Egypt in its sights.

A COMING WAR FOR THE WATERS OF THE NILE?
SOME ANCIENT PROPHECIES

Between the two old sayings *Africa semper Africa* and *Ex Africa semper aliquid novi* ('Africa will always be Africa' and 'Always something new in Africa'), it is to be feared that the second is more likely to prove to be true than the first.[16] In the long run the problem of the Nile fully justifies the statement in 1987 of Boutros Boutros-Ghali, Sadat's Minister for Foreign Affairs, that 'The next war in our region will be fought over water, not politics'. He is a leading Egyptian with a lucid view of the future.

The Africa of tomorrow will not be the Africa of yesterday. The twentieth century has seen the end of colonialism; in the course of the third millennium, with ups and downs, the Africans will gradually reshape their continent. They will break with the past and we shall see the emergence of new undertakings that are not under the thumb of others; states will claim all their rights and the peoples, seen by pessimists as without a future, will wake up politically and economically. Already Khartum is giving up a war that is costing it a million dollars a day and is from time to time offering independence to South Sudan.

The situation will be reversed and those distant lands on the banks of the Nile, where the Pharaohs came to collect their 'human merchandise' will see the Egyptians come to beg for water. The floor of Lake Nasser will be dredged and sluices will be pierced through the bottom of the high dam in order to let through a meagre stream of water. The turbines, and the Japanese pumps at Tushka, will have stopped long before.

Perhaps the prophecy of Ezekiel,[17] that priest exiled to Babylon who sought to preserve the hopes of the Chosen People during their captivity, will then come to pass: 'The Lord Yahweh says this: Now I set myself against you, Pharaoh king of Egypt, you great crocodile wallowing in your Niles, you have said, 'My Niles are mine, I made them'.... Thus saith the Lord God... the Nile is mine, I made it. I mean to reduce Egypt to desert and desolation, from Migdol to Syene and beyond to the frontiers of Ethiopia... and I am going to dry up the arms of the Nile, hand the country over to brigands, and lay the country waste and everything in it'. It is rather surprising to see Ezekiel using the plural for the Nile; he lived in the sixth century before our era, long after the two Niles had joined up, and was probably alluding to the many branches in the delta.

Isaiah is not more reassuring: 'The waters will ebb from the Nile, the river bed will be parched and dry, the canals grow foul, the Niles of Egypt sink and dry up, rush and reed will droop. The plants on the banks of the Nile, all the Nile vegetation will dry up, blow away, and be seen no more. The fishermen will groan, all who cast hook in the Nile will mourn; those who throw nets on the water will lament'. Here, at the start of the third millennium, the pollution of the Nile's waters, the sterility of certain plots in the cities, the disappearance of sardines from the delta and the regression of the delta give credence to the biblical prophet's words nearly three thousand years ago.

---

16. The flow of capital to African countries other than South Africa is rising for the fourth consecutive year.
17. Ezekiel 29, verses 3, 9 and 10, and 30, verse 12.

Human beings and the Nile lived their lives together: the Nile had shown *homo sapiens* the route north through the long plain where the world's first civilization blossomed. Once it was the cement of human societies, it will tomorrow become the fomenter of division. Since the golden age of the Pharaohs, whose gold-covered boats 'illumined the Nile', what changes there have been! Now deep shadows are spreading over the king of rivers, doomed to be the hapless cause of future wars.

'Father Terror' has had his eyes fixed on the Nile for thousands of years. The Greek word for 'Sphinx' means 'to grip tightly, to strangle'. He must be wondering anxiously if there really exist forces of evil that will succeed in strangling the waters of the generous river.

Paradoxically, for many Egyptians, the high dam has become like a tranquilizer: its sluice-gates will always be there ready to distribute the liquid so necessary to the country's prosperity. They forget that this great river is 'the distant one'[18] and that on its banks live those whom Ezekiel called 'brigands'.

Let us not be too pessimistic. We have a great love of Egypt through all its centuries and we hope that it will wake up to the dangers lying in wait for it, relinquish its risky projects and find the key to a problem that will continually evolve.

Just as President Mubarak likes to tell his country of the great opportunities opened up by the new valley, the President of Ethiopia assures his countrymen that the golden age will be that of irrigation and energy-generating waterfalls. But each of them is perfectly aware of the stubborn facts. The new valley was a breath of oxygen promised to his people on the eve of his fourth term of office and Ethiopia's golden age is not for tomorrow: the country is poor, and the World Bank has confirmed that it is not interested in the projects announced. Naturally, Egypt will do nothing immediately to change its policy, for its President is probably familiar with what Herodotus, after his journey to Asia, wrote about the 'gorges' or channels of the Great King: 'There is a plain in Asia surrounded by a ring of hills, which are broken by clefts in five separate places... but ever since the Persian rise to power it has been the property of the Persian king. In the ring of hills a considerable river arises.... The king has blocked up the gorges and constructed sluice-gates to contain the flow of water, so that what used to be a plain has now become a large lake, the river flowing in as before but no longer having any means of egress. The result of this for the people who depended upon the use of the water, but are now deprived of it, has been disastrous.... They go in a body to the Persians, and stand howling in front of the gates of the king's palace, until the king gives orders to open the sluices.... He opens the sluices only upon receipt of a heavy payment' (Herodotus, Histories, Book III, 117).

The site concerned is not Ethiopia, but there still exists in the plain of the lower

---

18. From the title of the book by Christiane Desroches Noblecourt, *Amours et Fureurs de la Lointaine* (Loves and Fury of the Distant One).

Nile the same distrust as the Pasha of Cairo used to feel towards the 'very Christian Negus'. The idea that the king of Ethiopia was master of the Nile and in a position to starve Egypt of water was current in Europe as long ago as the fourteenth century. Pilgrims to the Holy Land mention the fact and certain Ethiopian texts of the eighteenth century, such as the *maṣḥafa ṭêfut*,[19] lent credence to this belief. The latter text describes various conflicts between Muslims and the Coptic Patriarch who, under intense pressure, was saved by the king of Ethiopia. 'The water of the river that rises in my country will be stopped from reaching yours, which I shall cause to die of thirst', thundered the fourteenth-century Negus Dawit II. But exactly how would he do that? The answer is left to the imagination, but the king is generously credited with having the power to halt the course of the Blue Nile.

Nowadays, the short and medium term danger is more likely to come from the upriver countries of the White Nile, where pumping for irrigation will be boosted by the development of the oilfields. That will reduce drastically the flow of the Nile at low water. With the crisis in Sudan in recent years, the share of agriculture in the country's Gross Domestic Product has risen sharply.

A LONG-TERM SOLUTION: FEDERATIVE COOPERATION BETWEEN THE THREE BLUE NILE STATES

Time is working against the Egyptians: the upriver developing countries will gradually learn the techniques of irrigation, which will be highly productive but greedy for water. Masters of the plain of Egypt, do not display the arrogance of the Pharaohs! Be cooperative, for otherwise your country will die of thirst. A ray of hope: just after his re-election for a fourth term of office, President Mubarak modified his position. It is now said that major projects must be economically viable: what cannot be financed by the state must be financed by the private sector, the implication being that unprofitable projects will be abandoned. There are also signs of some important impending changes in Egypt's international relations, for President Mubarak is making advances towards his neighbour to the south, Omar al-Bashir, after a decade of being at loggerheads.

After a period of penury and incomprehension, the natural outcome will be an agreement between Egypt, North Sudan and Ethiopia. These three countries will join forces to build, sharing the expenses, the few dams needed in Ethiopia to tame the Blue Nile. The waters of the monsoon will be stored and released at the right moment. Ethiopia will serve as the water tower and source of electricity for the three countries, rather like an African Switzerland, forcing the generous but violent waters of the monsoon to drive generators throughout the year that will provide the countries down the river with the necessary energy for the development of their industries in return for food and industrial products.

---

19. A rare book that has been dated and translated with a commentary by André Caquot, professor at the Collège de France. See *Annales de l'Ethiopie*, 1955, fasc. I, pp. 89-108.

There will be no more inundations of the Nile but its low-water discharge will be much increased; the Blue Nile and the Atbara will no longer strip the slopes of the valleys to silt up Lake Nasser. The lake will then rediscover the taste of water, in the same way as the Aral Sea will when the necessary capital has been raised to finish a dam on the Syr Darya river.

The three countries will have to arrive at a perfect understanding with no hidden motives. When that time comes we shall see, like in ancient times under Narmer, the double land of Upper and Lower Egypt stretching from Aksum to Alexandria irrigated by a river with a generous and regular flow that will once again offer prosperity to a new form of cooperation between the three Blue Nile states.

Strange Africa, strange Nile. On the eastern mountain slopes of that mysterious continent is born the river that has inspired so many myths and given those living on its banks such intelligence, skill and ambition. The glory of the Pharaohs has not eclipsed the history of the Nile; indeed, it illuminated its finest period. The Pharaoh, master of the river, gave water to the land of the plain and today all the men and women living in the valley want to do the same. Under their assaults, the river will waste away and its peoples fly at one another's throats. When early man first set out northwards towards unknown lands, he was full of hope, but his numerous successors on the banks of the Nile now wonder whether its waters will be enough to slake their thirst. No other river in the world has aroused so much admiration, and so much concern for the future.

The hieroglyphs and hydraulic engineering

The scribes wrote for their fellow-countrymen and not for Egyptologists. The language they used borrowed a lot from everyday life and in particular from the hydraulic systems on which exchanges and communication in all parts of the plain were based. As the physical traces of many of these works have disappeared, we have to try to reconstitute the scene by shedding a little extra light on a language that expert philologists tell us is still far from being perfectly understood.

Professor Gérard Roquet, an Egyptologist and philologist, has authorized reproduction of the following excerpt from a letter to the author:

'The questions raised by your work on the canal and port will, in my opinion, have a crucial impact on the interpretation of the *Pyramid texts*, since the language used in that fundamental corpus hinges on the constant imagery of a lost but doubtless reclaimable landscape in which ancient hydraulic works and river transport are at the centre of textual problems that have hitherto been unnoticed, ignored or regarded as obscure.'

APPENDIX 2

The high dam, a challenge to the west

The idea of building the high dam took shape in Egypt in 1954. In October 1955 Ahmed Hussein, the Egyptian Ambassador in Washington informed the Department of State that the Russians had offered to finance the construction of a high dam , the preliminary plans for which had been drawn up by Egypt. The United States, afraid of being left behind as it had been over the sale of arms, stated its willingness to accord a substantial assistance. The cost of the entire scheme was estimated at 1.25 to 1.5 billion dollars. The press was strongly in favour: the *New York Herald Tribune* considered that the practical and symbolic significance of that constructive act was of vast political import and rendered the Soviet offer of assistance very dangerous.

As a gift, the United States offered 70 million dollars. Two months later the International Bank for Reconstruction and Development, after examining the preliminary plans, proposed a loan of 200 million dollars. The reactions of the Egyptian press to these proposals were embittered. On 2 April, 1956, Colonel Nasser told the *New York Times* correspondent that he had not rejected the offers of the Soviet Union and that he was keeping them in reserve should the discussions with the United States come to nothing.

In June 1956 there was a clearly hostile current among the Members of Congress and the Senate Budget Commission voted against the financial contribution. Nasser hesitated. Dr Kaissouni, his Minister for Finance, thought it worth reminding the West of the matter of the sale of arms.

The United States re-examined the question. The Secretary of State, John Foster Dulles, did not feel it necessary to bring it before the Senate: he was convinced that Nasser, realizing the danger for Egypt of too close relations with the Kremlin, would return to the Western fold. On 19 July, he convoked the Egyptian Ambassador to inform him that the United States Government had decided to withdraw its offer made the previous December for the financing of the high dam. Britain followed suit.

On 25 July Nasser, in inaugurating the Suez-Cairo oil pipeline, violently attacked the United States. A few days later, he received several times the Ambassador of the Soviet Union and concluded on 27 July a military assistance treaty. Feeling himself sufficiently protected, the Raïs nationalized the canal. The final plans for the dam were then drawn up by Hydroprojekt in Moscow, but all the implications of such a huge project, especially its excessive scale, were not considered. Although the river was becoming more moderate in its behaviour, it was represented – in the sort of flight of fancy to be expected just after a revolution – as an irate monster whose violent fits of temper would be successfully tamed. In a similar way, just after their revolution, the Chinese Republic let it be known that it would master the fury of the Dragon (the Yangtse-kiang) by damming its course at the Three Gorges – a much more risky undertaking than that of Nasser.

Use of the grand gallery as a hoist:
A few calculations

With seven men per tonne to be pulled, a sledge weighing a total of 60 tonnes could, with the traditional silt lubrification, mount a construction ramp with a 5% incline at the low speed of 0.30 metres per second[1]. It would therefore require $7 \times 60 = 420$ hauliers , whose efforts would be difficult to coordinate.

If the sledge were attached to a rope pulled by blocks of stone descending the Grand Gallery, the work of the hauliers would be considerably facilitated. By how much?

The German archaeologist Borckhard has represented the Grand Gallery with the wooden floor on which one of the plugs-blocks that would later serve to conceal the access corridor and under which the procession of priests would pass to go to the King's Chamber, was kept. These blocks were cut to within a centimetre of the height and width of the corridor. They weighed five tonnes. Let us assume first of all that three of them were attached together for the descent along the lubricated floor.

Their motive power is:

$$3 \times 5 \times \sin 26°56 = 6.70 \text{ t}$$

(26°56 corresponds to the angle of slope of the Grand Gallery). From this must be deducted the friction component which, assuming that the wooden floor was lubricated with slightly warmed animal fat, would be 0.04. Hence

$$3 \times 5 \times \cos 26°56 \times 0.04 = 0.54 \text{ t}$$

This leaves 6.16 t, a figure that must be further reduced to allow for the friction of the rope sliding on a greased block:

$$\exp 0.04 \times 0.51 = 1.02$$

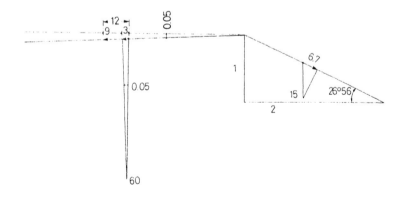

---

1. J. Kerisel, *op. cit.*, p. 210.

where 0.04 represents the friction and 0.51 the angle of deflection, in radians, of the rope. There therefore remains for the hauling of the sledge 6.16 divided by 1.02 = 6.04 t.

If now we estimate the force needed to pull a sledge weighing 60 t, we find by the same means:

$$60 \times 0.05 + 60 \times 0.99 \times 0.15 = 12 \text{ t}$$

(0.05 being the slope and 0.15 the friction).

The lowering of three five-tonne blocks into the Grand Gallery therefore made it possible to do without half the hauliers and thus coordinate the operation more easily.

The two hundred men thus freed would have been used to pull up, one by one, the stone plugs. This would require a force equal to a third of the total 6.70 + 0.54 = 7.24 t, i.e. 2.41 t, which, allowing a tractive effort of 12 kg for each man, would require 200 men. The 400 men would there be divided into two teams working one after the other with improved coordination.

Note also that the use of six blocks attached together would pull up one of the stone slabs without assistance, as shown in the text.

Index

For Product Safety Concerns and Information please contact our EU
representative GPSR@taylorandfrancis.com
Taylor & Francis Verlag GmbH, Kaufingerstraße 24, 80331 München, Germany

www.ingramcontent.com/pod-product-compliance
Ingram Content Group UK Ltd.
Pitfield, Milton Keynes, MK11 3LW, UK
UKHW051827180425
457613UK00007B/239

9 789058 093431